T0184672

BestMasters

Maximilian Scheurer

Polarizable Embedding for the Algebraic-Diagrammatic Construction Scheme

Investigating Photoexcitations in Large Biomolecular Systems

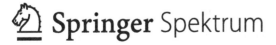

Springer Spektrum

Maximilian Scheurer
Heidelberg, Germany

ISSN 2625-3577   ISSN 2625-3615   (electronic)
BestMasters
ISBN 978-3-658-31280-0   ISBN 978-3-658-31281-7   (eBook)
https://doi.org/10.1007/978-3-658-31281-7

This Springer Spektrum imprint is published by the registered company Springer Fachmedien Wiesbaden GmbH part of Springer Nature.
The registered company address is: Abraham-Lincoln-Str. 46, 65189 Wiesbaden, Germany

Abstract

Light-driven biochemical reactions are ubiquitous and essential in all living organisms. Such photo-induced processes take place in active sites of enzymes that are highly optimized with respect to their unique structure providing necessary catalysis in the reaction mechanism. Often, the electronic excitation process can be approximately confined to the chromophore embedded in the protein. Such chromophores are not unique to a specific protein but are found in a multitude of different proteins with different cellular functionality. As a matter of fact, the spectral properties of chromophores strongly depend on the protein environment. In the last decades, such photoactive proteins and enzymes have been studied not only in experiments but also through computational methods. In this context, it is often prohibitive to study these systems in their entirety by means of quantum chemical *ab-initio* calculations. However, as explained above, effects of the protein environment can be essential to elucidate spectral properties, e.g., for computational mutant design to fine-tune light activation in optogenetics. To this extent, a novel combined method of the algebraic-diagrammatic construction (ADC) scheme with the polarizable embedding (PE) model is presented in this thesis. The aim of the PE-ADC method is to allow for highly precise computations of electronic excitations on a chromophore through ADC and integration of environment effects through PE. Essentially, the combined variant relies on a self-consistent Hartree-Fock (HF) ground state calculation in presence of the environment, a subsequent ADC calculation, and *a posteriori* corrections of the excitation energies. All the necessary methodology for inclusion of the polarizable environment has been implemented from scratch in a novel C++-based library, namely CPPE, which was fully integrated into the *Q-Chem* program package. The flexibility of the approach allows for ADC calculations through second and third order, and for usage of any ADC 'flavor'. Accuracy and large-scale applicability of the implemented method are demonstrated in three case studies. One of the case studies examines the photo-induced charge-transfer (CT) excitation which is key to the photoprotection mechanism in the dodecin protein. Hence, this thesis presents a comprehensive elaboration on the new PE-ADC method, comprising the theoretical model, its implementation, and state-of-the-art application to a photo-biochemical process.

Abbreviations

ADC algebraic-diagrammatic construction

AO atomic orbital

CC coupled cluster

COSMO conductor-like screening model

CSM continuum solvation model

DFT density-functional theory

FDE frozen density embedding

FQ fluctuating charges

GIAO gauge-invariant atomic orbital

HF Hartree-Fock

MD molecular dynamics

MM molecular mechanics

NTO natural transition orbital

ON occupation number

PCM polarizable continuum model

PDE polarizable density embedding

PE polarizable embedding

QM quantum-mechanical

TDDFT time-dependent density-functional theory

SCF self-consistent field

SQ second quantization

Contents

1 Introduction

A plethora of ubiquitous biological processes, such as photosynthesis, phototropism, phototaxis and vision are driven by the interaction of light and matter, i.e., by electronic excitations[1]. In this context, computational approaches can assist to understand the underlying physicochemical mechanism on an atomistic scale. In recent years, high-level *ab-initio* quantum-mechanical (QM) calculations of large and biologically relevant molecules have become increasingly popular[2–7]. This is, on one hand, driven by the steady increase in computational resources and, on the other hand, by improved algorithms. In addition, substantial advances in environmental embedding schemes have enabled a multi-scale treatment of large systems[2,6–8]. A multitude of models has emerged aiming at the description of molecular systems for which a full QM description is not needed but environmental effects are non-negligible. Implicit embedding schemes mimic the medium surrounding a specific molecule by a dielectric continuum. For these continuum solvation models (CSMs), multiple formulations have evolved, e.g., the polarizable continuum model (PCM) and the conductor-like screening model (COSMO)[9–12]. Calculations involving CSMs usually aim at the description of a liquid solvent and are straightforward to perform since only a dielectric constant and the cavity need to be chosen. One of the most severe shortcomings of CSMs lies in modeling directionality of explicit solvent interactions, such as hydrogen bonds or π-π-stacking[13,14]. These interactions are, however, crucial in anisotropic, heterogeneous environments, such as protic solvents, proteins, nucleic acids and biological membranes. Explicit embedding schemes, on the contrary, capture these heterogeneities by retaining a full atomistic representation of the environment[2]. As a result, explicit quantum-classical embedding schemes are generally

© The Editor(s) (if applicable) and The Author(s), under exclusive license to
Springer Fachmedien Wiesbaden GmbH, part of Springer Nature 2020
M. Scheurer, *Polarizable Embedding for the Algebraic- Diagrammatic
Construction Scheme*, BestMasters, https://doi.org/10.1007/978-3-658-31281-7_1

more complex and computationally more demanding than CSMs as they require parametrization of the atomistic environment. In the simplest case, the parameters can be obtained by directly transferring charges from commonly available molecular mechanics (MM) force fields. However, such parameters are rarely intended to reproduce the microscopic detail required for accurate embedding calculations, especially with respect to molecular response properties. For the latter, not only a description of permanent electrostatics is of great importance, but also mutual polarization effects between the environment and the core region are desired. It is important to recognize that the increase in complexity and computational effort of explicit embedding arises not only from the embedding as such, but also from the discrete nature of the environment which requires sampling of a representative number of configurations.

The polarizable embedding (PE) model[15,16] is a fragment-based, hybrid quantum-classical, and explicit embedding scheme, capable of modeling the aforementioned physical effects. Parameters are obtained from first-principles QM calculations performed on small split fragments of the molecular environment. Hence, close agreement with full QM calculations can in general be expected if the property of interest is localized on the quantum part only. In PE, the permanent charge distribution of the environment is modeled by fragment-based multi-center multipole expansions through a specific order, usually including monopole (charges), dipole, and quadrupole moments centered at each atom. Polarization effects are taken into account by dipole-dipole polarizabilities placed at the multipole expansion sites. Thus, PE is capable of treating mutual polarization of the environment and the quantum region in a self-consistent manner. In particular, PE has been designed to obtain accurate molecular response properties, e.g., electronic excitation energies. Accordingly, it has already been combined with several methods for calculating excited states in solution, i.e., time-dependent density functional theory (TD-DFT)[15–17], coupled-cluster (CC) methods[18–20], second-order polarization propagator (SOPPA)[21] as well as a formulation within resonant-convergent response theory[22] and an open-ended response theory framework[23].

Further, these combined approaches have been successfully applied to study biomolecules, e.g., probes for biological membranes [24,25], nucleic acids [26], fluorescent proteins [27–29] and light-gated ion channels [30]. For a recent perspective on the PE model see Ref. [31].

The choice of electronic structure method is decisive for the quality of the excited state description. Especially TD-DFT with standard exchange-correlation functionals has well-known limitations in the treatment of charge-transfer (CT) excitations, doubly excited states and Rydberg states [32–35]. In recent years, the algebraic-diagrammatic construction (ADC) scheme has emerged as an accurate and balanced method for the calculation of excited states [36]. It has also been extended to study core-excitation [37,38] and ionization processes. The ADC method builds on a Hermitian representation of the Hamiltonian, is size-consistent and systematically improvable. Also, the computational effort of ADC through second order (ADC(2)) is still manageable for medium-sized systems [36]. ADC through third order (ADC(3)) is particularly accurate in treating electronically doubly excited states [36,39]. The ADC method was previously combined with several embedding schemes, such as PCM [40,41], COSMO [42,43], and frozen density embedding (FDE) [44,45]. Until recently, there was no combination of ADC with a hybrid quantum-classical, explicit embedding scheme including a polarization treatment for modeling electronic excitations in almost arbitrarily complex and large environments.

In this thesis, the first combination of PE with ADC is developed, implemented, tested and applied to a biologically relevant system. The PE-ADC scheme is in principle independent of the ADC 'flavor' as it is built upon a PE-HF ground state calculation and takes into account perturbative corrections of the excitation energies in a density-driven manner. That is, as long as transition densities and state densities are available for the ADC variant, PE-ADC is immediately feasible. At the same time that this thesis and the corresponding publication were finished, Hättig and co-workers published a similar approach including analytic excited states gradients for PE-ADC(2) [46]. Their approach, in contrast to the here presented methodology, includes the environment response in the ADC calculation. However, only ADC(2) is available and the work has obvious flaws with the environment parametrization. As the results on test systems will show, inclusion of the environment response in the ADC calculation is not necessarily the better way to go, given the increased computational cost that comes with this approach.

The thesis is structured as follows: First, the necessary theoretical methodology is briefly outlined. Second, the theoretical basis for the density-driven combination of PE and ADC is established. Afterwards, I thoroughly describe the implementation of a novel library for PE calculations, CPPE, which is one of the cornerstones of this work. Furthermore, the computational methodology of all performed calculations is presented. Thereafter, results on three test systems using PE-ADC are shown, including para-nitroaniline (PNA) in presence of small water clusters, the lumiflavin (Lf) chromophore in bulk solvent, and finally the 'flagship' application investigating the flavin-binding protein dodecin. The results obtained for dodecin extend previous theoretical work[47] by a key finding: The polarizable protein environment assists in stabilizing an energetically low-lying CT excitation which is assumed to trigger for the photoprotection mechanism in dodecin. Finally, a brief conclusion including a perspective on the novel method is given, including an outlook on ongoing and future projects that employ this thesis as necessary groundwork.

2 Theoretical Aspects

2.1 Second Quantization

The standard formulation of quantum mechanics as derived from the five postulates represent physical observables, e.g., energy, momentum, and position by means of operators. Solving the Schödinger equation yields a wave function which contains all the necessary information about the states of a system. Going one step beyond this formulation in *first quantization, second quantization* (SQ) also expresses wave functions, i.e., states by operators acting on the vacuum state[48]. The elementary operators acting on the vacuum state are the so-called creation and annihilation operators. From the properties of fermionic wave functions, one can derive all the algebraic relationships of the elementary operators. As follows, the first quantization operators can also be expressed in terms of the elementary operators, resulting in a unified formalism to construct both operators and wave functions. Another advantage of second quantization (SQ) is that all equations can be written in a particle-number independent manner, i.e., there is never an explicit reference to a *specific* electronic coordinate. Based on Ref. 48, I will briefly outline the important mathematical aspect of SQ together with operator representation as the PE model was formulated in this framework.

2.1.1 Fock Space

Let us consider a basis of spin orbitals $\{\phi_p(\underline{\mathbf{r}}, \sigma)\}$ with the spatial coordinate $\underline{\mathbf{r}}$ and the spin coordinate σ. To construct trial functions for multi-electron systems, one employs so-called *Slater determinants* which are nothing but anti-symmetrized products of multiple spin

© The Editor(s) (if applicable) and The Author(s), under exclusive license to
Springer Fachmedien Wiesbaden GmbH, part of Springer Nature 2020
M. Scheurer, *Polarizable Embedding for the Algebraic- Diagrammatic
Construction Scheme*, BestMasters, https://doi.org/10.1007/978-3-658-31281-7_2

orbitals. We now define a vector space, the *Fock space*, to represent Slater determinants in an abstract manner. Each determinant is then described by an occupation number (ON) vector $|\mathbf{k}\rangle$ given by

$$|\mathbf{k}\rangle = |k_1, k_2, \ldots, k_m\rangle, \, k_p = \begin{cases} 1 & \phi_p \text{ occupied} \\ 0 & \phi_p \text{ unoccupied} \end{cases} \qquad (2.1)$$

The occupation number k_p is 1 if the spin orbital ϕ_p is present in the Slater determinant, and 0 if the spin orbital is not present. The scalar product of two ON vectors $|\mathbf{k}\rangle$ and $|\mathbf{m}\rangle$ for a set of N orthonormal spin orbitals is simply

$$\langle \mathbf{k} | \mathbf{m} \rangle = \prod_{p=1}^{N} \delta_{k_p m_p}. \qquad (2.2)$$

Actually, this inner product is also well-defined for states with different particle numbers. In that case, the overlap is zero.

2.1.2 Elementary Operators

Now, we will elaborate on the elementary creation and annihilation operators that will be used to construct states and operators. The *creation* operator, $\hat{a}_p^\dagger$, creates an electron in spin orbital p, i.e.,

$$\hat{a}_p^\dagger |k_1, k_2, \ldots, 0_p, \ldots, k_m\rangle = \Gamma_p^{\mathbf{k}} |k_1, k_2, \ldots, 1_p, \ldots, k_m\rangle. \qquad (2.3)$$

The phase factor $\Gamma_p^{\mathbf{k}}$ is needed to obtain operators consistent with first quantization and is either -1, if there is an odd number of spin orbitals on the left-hand side of k_p or $+1$, if there is an even number of spin orbitals in that range. If the vector already contains an electron in spin orbital p, i.e., $k_p = 1$, action of $\hat{a}_p^\dagger$ on this determinant gives zero,

$$\hat{a}_p^\dagger |k_1, k_2, \ldots, 1_p, \ldots, k_m\rangle = 0 \qquad (2.4)$$

This represents the usual structure of a Slater determinant which is also zero if a spin orbital occurs more than once. The respective relations for the annihilation operator, $\hat{a}_p$, are given by

$$\hat{a}_p \left| \mathbf{k} \right\rangle = \delta_{k_p 1} \Gamma_p^{\mathbf{k}} \left| k_1, k_2, \ldots, 0_p, \ldots, k_m \right\rangle . \qquad (2.5)$$

Thus, the operator reduces the occupation number k_p from 1 to 0 if spin orbital p is occupied, whereas it produces 0 if the spin orbital is not occupied. Furthermore, one can derive anticommutation relations for the elementary operators[48], here listed for completeness:

$$\{\hat{a}_p^\dagger, \hat{a}_q^\dagger\} = 0 \qquad (2.6)$$

$$\{\hat{a}_p, \hat{a}_q\} = 0 \qquad (2.7)$$

$$\{\hat{a}_p^\dagger, \hat{a}_q\} = \delta_{pq} \qquad (2.8)$$

From the anticommutation relations, all subsequent algebraic properties of SQ can be formulated.

2.1.3 Operators in Second Quantization

Operators in first quantization depend on the number of electrons in the molecular system and have a spatial structure. Until now, we have not introduced any spatial structure to the SQ formalism. Hence, to comply with first-quantized operators, the SQ operators must contain a spatial structure as well. One can show that a one-electron operator $\hat{v}^c$ given by

$$\hat{v}^c = \sum_{i=1}^{N} v^c(\underline{\mathbf{r}}_i, \sigma_i) \qquad (2.9)$$

corresponds to

$$\hat{v} = \sum_{pq} v_{pq} \hat{a}_p^\dagger \hat{a}_q \qquad (2.10)$$

in second quantization. The operator in SQ does not depend on the electronic coordinates explicitly, but on the integrals over the operator in the given spin orbital basis, i.e.,

$$v_{pq} = \int \phi_p^*(\underline{x})v^c(\underline{x})\phi_q(\underline{x})\mathrm{d}\underline{x} \qquad (2.11)$$

which also introduces the spatial structure of the operators. Here, $\underline{x}$ is a collective variable of spatial and spin coordinates. In case the one-electron operator acts on spatial coordinates only, one can also formulate the operator by summing over spin coordinates as

$$\hat{v} = \sum_{pq} v_{pq}\hat{E}_{pq} \qquad (2.12)$$

with

$$\hat{E}_{pq} = \hat{a}_{p\sigma}^\dagger \hat{a}_{q\sigma} + \hat{a}_{p\tau}^\dagger \hat{a}_{q\tau}. \qquad (2.13)$$

Here, $\hat{E}_{pq}$ is the orbital singlet excitation operator. Similarly, one can derive the SQ formulation of a two-electron operator as

$$\hat{g} = \sum_{pqrs} g_{pqrs}\hat{a}_p^\dagger \hat{a}_r^\dagger \hat{a}_s \hat{a}_q \qquad (2.14)$$

with the four-index two-electron integrals

$$g_{pqrs} = \iint \phi_p^*(\underline{x_1})\phi_r^*(\underline{x_2})g^c(\underline{x_1}, \underline{x_2})\phi_q(\underline{x_1})\phi_s(\underline{x_2})\mathrm{d}\underline{x}_1 \mathrm{d}\underline{x}_2. \qquad (2.15)$$

Summing over spin variables, we can also write

$$\hat{g} = \sum_{pqrs} g_{pqrs}\hat{e}_{pqrs} \qquad (2.16)$$

with the two-electron excitation operator

$$\hat{e}_{pqrs} = \hat{E}_{pq}\hat{E}_{rs} - \delta_{qr}\hat{E}_{ps} \qquad (2.17)$$

Using these expressions for one- and two-electron operators, we can in the following section formulate the electronic Hamiltonian operator and make use of a special wave function parametrization in second quantization.

2.2 Hartree-Fock Theory

Using the SQ formalism, the electronic Hamiltonian is given by

$$\hat{H} = \sum_{pq} h_{pq} \hat{E}_{pq} + \frac{1}{2} \sum_{pqrs} g_{pqrs} \hat{e}_{pqrs} + V_{\text{nuc}}, \qquad (2.18)$$

The integrals are given by

$$h_{pq} = -\frac{1}{2} \int \phi_p^*(\underline{r}) \nabla^2 \phi_q(\underline{r}) \mathrm{d}\underline{r} - \sum_n^M \int \phi_p^*(\underline{r}) \frac{Z_n}{|\underline{r} - \underline{R}_n|} \phi_q(\underline{r}) \mathrm{d}\underline{r} \quad (2.19)$$

and

$$g_{pqrs} = \iint \phi_p^*(\underline{r}) \phi_r^*(\underline{r}') \frac{1}{|\underline{r} - \underline{r}'|} \phi_q(\underline{r}) \phi_s(\underline{r}') \mathrm{d}\underline{r}\mathrm{d}\underline{r}'. \qquad (2.20)$$

Here, h_{pq} denotes the integrals over the kinetic and nuclear attraction Coulomb operator (*core* Hamiltonian) whereas g_{pqrs} are the two-electron Coulomb repulsion integrals. Further, V_{nuc} is the nuclear repulsion energy of the M nuclei of the molecule, given by

$$V_{\text{nuc}} = \sum_{n<m}^N \frac{Z_n Z_m}{|\underline{R}_n - \underline{R}_m|}. \qquad (2.21)$$

In Hartree-Fock (HF) theory, one assumes that the N-body wave function of the electronic ground state can be represented by a single Slater determinant that minimizes the energy among all possible choices of Slater determinants. In the following, we will consider a closed-shell state $|\text{cs}\rangle$, i.e., the orbitals are either doubly occupied or unoccupied and referred to as *inactive* and *virtual* orbitals, respectively. I will use the indices i, j, k, l for occupied orbitals, a, b, c, d for virtual orbitals, and p, q, r, s as general orbital indices.

As stated above, one wants to find the wave function that minimizes the energy

$$E = \langle \text{cs}|\hat{H}|\text{cs}\rangle. \qquad (2.22)$$

In the SQ formalism, one can choose a parametrization of the wave function using the elementary operators. Note that the closed-shell state is given by

$$|cs\rangle = \left(\prod_{p}^{N} \hat{a}_{p\alpha}^{\dagger} \hat{a}_{p\beta}^{\dagger} \right) |vac\rangle . \qquad (2.23)$$

From a given trial wave function $|cs\rangle$, one can determine a new wave function by a unitary transformation as

$$\widetilde{|cs\rangle} = e^{-\hat{\kappa}} |cs\rangle \qquad (2.24)$$

using

$$\hat{\kappa} = \sum_{pq} \kappa_{pq} \hat{E}_{pq}. \qquad (2.25)$$

The operator $\hat{\kappa}$ is anti-Hermitian ($\hat{\kappa}^{\dagger} = -\hat{\kappa}$), such that its exponential is a unitary operator, i.e., one need not take care of proper wave function normalization since this is already inherently contained in the parametrization. As a unitary transformation corresponds to a rotation in three dimensions, the exponential operator *rotates* the orbitals, and κ_{pq} are the corresponding orbital rotation parameters. Let us now determine which rotations among orbitals are actually non-redundant. It is trivial to see that

$$\hat{E}_{ab} |cs\rangle = 0, \qquad (2.26)$$

and further

$$\hat{E}_{ij} |cs\rangle = 0. \qquad (2.27)$$

That is, rotations among inactive or virtual orbitals themselves are redundant and do not contribute. Only orbital rotations between inactive and virtual orbitals are non-redundant, i.e.,

$$\hat{\kappa} = \sum_{ai} \kappa_{ai} \hat{E}_{ai}. \qquad (2.28)$$

With this wave function ansatz, the energy subject to minimization depends on the orbital rotation parameters, i.e., $E = E(\hat{\boldsymbol{\kappa}})$. We can now write the stationary condition of the energy as

$$\frac{\partial}{\partial \kappa_{pq}} \langle \mathrm{cs}|e^{\hat{\boldsymbol{\kappa}}} \hat{H} e^{-\hat{\boldsymbol{\kappa}}}|\mathrm{cs}\rangle \Bigg|_{\hat{\boldsymbol{\kappa}}=0} = 0, \qquad (2.29)$$

that is, the energy must be stationary with respect to small rotations in the orbitals. The energy can be written by a Baker-Campbell-Haussdorff (BCH) expansion, yielding

$$E(\hat{\boldsymbol{\kappa}}) = \langle \mathrm{cs}|\hat{H}|\mathrm{cs}\rangle + \langle \mathrm{cs}|[\hat{\boldsymbol{\kappa}}, \hat{H}]|\mathrm{cs}\rangle + \langle \mathrm{cs}|[\hat{\boldsymbol{\kappa}}, [\hat{\boldsymbol{\kappa}}, \hat{H}]]|\mathrm{cs}\rangle + \ldots . \qquad (2.30)$$

Using the stationary condition from above, we find

$$\frac{\partial}{\partial \kappa_{pq}} E(\hat{\boldsymbol{\kappa}}) \Bigg|_{\hat{\boldsymbol{\kappa}}=0} = \langle \mathrm{cs}|[\hat{E}_{pq}, \hat{H}]|\mathrm{cs}\rangle + \frac{1}{2} (1 + P_{pq,rs}) \langle \mathrm{cs}|[\hat{E}_{pq}, [\hat{E}_{rs}, \hat{H}]]|\mathrm{cs}\rangle$$
$$+ \ldots, \qquad (2.31)$$

where the first-order and second-order terms are the so called *electronic gradient* and *electronic Hessian*, respectively. Hence, the first-order condition requires an electronic gradient equal to zero, i.e.,

$$\langle \mathrm{cs}|[\hat{E}_{ai}, \hat{H}]|\mathrm{cs}\rangle = 0. \qquad (2.32)$$

One can immediately see that this is nothing but *Brillouin's theorem* stating that the electronic ground state does not couple to singly excited determinants, i.e., the matrix elements of the Hamiltonian between such determinants are zero. From this condition, one can derive an effective one-electron operator, the *Fock operator* $\hat{\mathcal{V}}_{\mathrm{HF}}$, given by

$$\hat{\mathcal{V}}_{\mathrm{HF}} = \sum_{pq} h_{pq} \hat{E}_{pq} + \sum_{pqi} (2g_{pqii} - g_{piiq}) \hat{E}_{pq}. \qquad (2.33)$$

The one-electron part of the Fock operator is equal to the core Hamiltonian, giving rise to the fact that the Fock operator matches the exact

Hamiltonian in a non-interacting case. Consequently, the second term describes interactions between electrons. Of note, each electron moves in a mean field created by all the other electrons. This is the result from the single-determinant ansatz since no other assumptions or approximations have been made in HF theory. For this reason, the Fock operator itself depends on the orbitals, i.e., the solution, in a non-linear manner. For this reason, the HF problem must be iteratively solved until the field is self-consistent. Such procedures are called self-consistent field (SCF) methods. In this context, the spin orbital basis is usually referred to as molecular orbitals (MOs). From a set of guess MOs, one constructs the Fock operator and diagonalizes it. The eigenvectors of the Fock operators are the so-called canonical MOs. From these newly obtained MOs, the Fock operator is constructed and again diagonalized until the newly obtained MOs match the ones that were used to build the Fock operator, i.e., until the field is self-consistent. Mathematically, the obtained $|cs\rangle$ determinant is an eigenfunction of the Fock operator, i.e.,

$$\hat{\mathcal{V}}_{\mathrm{HF}} \left|cs\right\rangle = 2 \sum_p \epsilon_p \left|cs\right\rangle, \qquad (2.34)$$

where the orbital energies ϵ_p are the eigenvalues of the Fock operator. These energies are given by

$$\epsilon_p = h_{pp} + \sum_i \left(2g_{ppii} - g_{piip}\right). \qquad (2.35)$$

Note that the HF energy is *not* the sum of orbital energies, but the expectation value of the Hamiltonian using the obtained Slater determinant, that is

$$\mathcal{E}_{\mathrm{HF}} = \langle cs|\hat{H}|cs\rangle = 2 \sum_i h_{ii} + \sum ij \left(2g_{iijj} - g_{ijji}\right) + V_{\mathrm{nuc}}. \qquad (2.36)$$

In the two-electron part, the first energy contribution results from Coulomb repulsion of the electrons, whereas the second part is the so-called exchange energy contribution which does not have a classical analogon. It accounts for the Fermi correlation of the electrons and is a result from the anti-symmetric wave function ansatz for fermions.

In practice, the HF equations are not solved directly. Instead, one makes use of the linear combination of atomic orbitals (LCAO) approximation. In this context, the MOs are expressed by linear combinations of atomic orbitals (AOs) as

$$\phi_p(\underline{\mathbf{r}}) = \sum_\mu C_{\mu p} \chi_\mu(\underline{\mathbf{r}}). \tag{2.37}$$

The atomic orbitals are represented by $\chi_\mu(\underline{\mathbf{r}})$ and the matrix $C_{\mu p}$ denotes the MO coefficients that build the molecular orbitals from the atomic orbitals. Subsequently, one can recast the HF equations to the AO basis such that the determined parameters are the MO coefficients. This gives rise to the *Roothaan-Hall* equations expressed in matrix form as

$$\mathbf{FC} = \mathbf{SC}\epsilon, \tag{2.38}$$

with the MO coefficient matrix $\mathbf{C}$ and the overlap matrix of atomic orbitals $\mathbf{S}$, given by

$$S_{\mu\nu} = \int \chi_\mu(\underline{\mathbf{r}})\chi_\nu(\underline{\mathbf{r}})\mathrm{d}\underline{\mathbf{r}}, \tag{2.39}$$

and $\mathbf{F}$ is the Fock matrix in AO basis, i.e.,

$$f_{\mu\nu} = h_{\mu\nu} + \sum_{\gamma\delta} P_{\gamma\delta}\left(g_{\mu\nu\gamma\delta} - \frac{1}{2}g_{\mu\delta\gamma\nu}\right). \tag{2.40}$$

In the last formula, the one-electron density matrix in AO basis $P_{\mu\nu}$ was used. The matrix elements are defined as

$$P_{\mu\nu} = 2\sum_i C_{\mu i}C_{\nu i}. \tag{2.41}$$

In a Roothan SCF procedure, the HF equations are solved until self-consistency. There are, of course, multiple ways of improving convergence of an SCF as well as analyzing convergence and stability of the solutions. These will be omitted here for simplicity. A code example of an SCF implementation can be found here: `https://github.com/maxscheurer/QMax`. Note that any external one-electron potential can be easily added to the Fock operator and thus to the SCF procedure. This will be of importance in deriving the effective operators for the PE model. In the following section, I will briefly outline how accurate excited states can be obtained, including explicit electron correlation and not only mean-field effects.

2.3 Algebraic-Diagrammatic Construction

The algebraic-diagrammatic construction of the polarization propagator (ADC) is a method to obtain *ab-initio* correlated excited states. A comprehensive derivation of ADC can be found in literature[36,49–51]. In brief, the intermediate state (IS) formalism is used to obtain the ADC working equations. A set of creation and annihilation operators is applied to the exact ground state wave function, and the set of resulting IS wave functions is subject to a Gram-Schmidt orthogonalization. Subsequently, the exact excited wave function $|\Psi_n\rangle$ is expanded in the basis of intermediate states, $\{|\tilde{\Psi}_J\rangle\}$, as

$$|\Psi_n\rangle = \sum_J X_{nJ} |\tilde{\Psi}_J\rangle , \qquad (2.42)$$

with the expansion coefficients X_{nJ} given through the eigenvectors of the ADC matrix. The ADC matrix, M_{IJ}, is a matrix representation of the Hamiltonian operator $\hat{H}$ shifted by the exact ground state energy E_0 in the basis of intermediate states, i.e.,

$$M_{IJ} = \langle \tilde{\Psi}_I | \hat{H} - E_0 | \tilde{\Psi}_J \rangle . \qquad (2.43)$$

The corresponding Hermitian eigenvalue problem reads

$$\mathbf{MX} = \mathbf{\Omega X} , \quad \mathbf{X}^\dagger \mathbf{X} = \mathbf{1}, \qquad (2.44)$$

with the eigenvector matrix $\mathbf{X}$ and the diagonal matrix $\mathbf{\Omega}$ containing excitation energies, i.e., $\Omega_n = E_n - E_0$, where n denotes the n-th excited state. In practice, the IS basis is constructed from a Møller-Plesset (MP) perturbation expansion of the Hartree-Fock (HF) ground state. This leads to a perturbation expansion of the ADC matrix, In this manner, MP(2) and MP(3) ground states yield ADC(2) and ADC(3) working equations, respectively. Explicit equations and implementations can be found in previous work[39,49–52]. I will briefly outline how ADC transition density matrices and excited state density matrices are obtained[50,51] because these are key to the density-driven approach to calculate *a posteriori* corrections. The one-particle reduced transition density matrix is given by

$$P_{pq}^{\mathrm{tr},0 \to n} = \langle \Psi_n | \hat{a}_p^\dagger \hat{a}_q | \Psi_0 \rangle = \sum_I X_{n,I}^\dagger \langle \tilde{\Psi}_I | \hat{a}_p^\dagger \hat{a}_q | \Psi_0 \rangle . \qquad (2.45)$$

Further, one constructs the one-particle reduced density matrices and the one-particle reduced transition density matrices between different excited states by

$$P_{pq}^{n \to m} = \langle \Psi_m | \hat{a}_p^\dagger \hat{a}_q | \Psi_n \rangle = \sum_{I,J} X_{m,I}^\dagger \langle \tilde{\Psi}_I | \hat{a}_p^\dagger \hat{a}_q | \tilde{\Psi}_J \rangle X_{n,J}. \qquad (2.46)$$

By setting $n = m$ in the above equation, one obtains the n-th excited state density matrix. The difference density matrix between the ground state and the n-th excited state is simply

$$\Delta \mathbf{P}^{0 \to n} = \mathbf{P}^n - \mathbf{P}^0. \qquad (2.47)$$

with the electronic ground state density $\mathbf{P}^0$ and the density of the n-th excited state $\mathbf{P}^n$. Employing the IS representation (ISR), it is straightforward to compute the density matrices from the ADC eigenvectors.

2.4 Polarizable Embedding

The polarizable embedding (PE) model is a fragment-based, hybrid quantum-classical and *explicit* embedding scheme. It aims at the description of electrostatics and polarization effects on a quantum-mechanically treated molecule through a classical parametrization of the environment. The permanent charge distribution of the environment is modeled by fragment-based multi-center multipole expansions, whereas polarization effecs are taken into account by dipole-dipole polarizabilities placed at the multipole expansion sites. Hence, mutual polarization of the environment and the core is treated self-consistently. In particular, PE has been designed to obtain accurate molecular response properties, e.g., electronic excitation energies.

I will, in the following, perform a full derivation of the PE model from scratch based on work by Kongsted and co-workers[15,16] and outline the approximations yielding the final working equations.

The objective of the PE model is to incorporate effects of the environment into the electronic density of a central molecular core system. This is, in the end, achieved by the PE operator.

2.4.1 Prerequisites

Consider a simple super-molecular system split up into two fragments A and B that consist of M_A and M_B nuclei, respectively.

The super-molecular Hamiltonian of the system is given by

$$\hat{H} = \underbrace{\hat{H}^A + \hat{H}^B}_{\hat{H}^{(0)}} + \hat{V}^{AB}, \tag{2.48}$$

with the fragment Hamiltonians $\hat{H}^A$ and $\hat{H}^B$, and the interaction Hamiltonian $\hat{V}^{AB}$. The sum of fragment Hamiltonians, i.e., the unperturbed Hamiltonian is denoted by $\hat{H}^{(0)}$.

Using the SQ formalism, the individual fragment Hamiltonian is simply

$$\hat{H}^i = \sum_{pq} h_{pq}^i \hat{E}_{pq}^i + \frac{1}{2} \sum_{pqrs} g_{pqrs}^i \hat{e}_{pqrs}^i + V_{\text{nuc}}^i, \tag{2.49}$$

with $i = A, B$. Note that the fragment Hamiltonian only contains terms specific to one fragment i. Here, h_{pq} denotes the integrals over the kinetic energy and nuclear-electron attraction operators, g_{pqrs} is the integral over the electron-electron repulsion operator, and V_{nuc} is the nuclear-nuclear repulsion energy. The integrals are defined as follows:

$$h_{pq} = -\frac{1}{2} \int \phi_p^*(\underline{r}) \nabla^2 \phi_q(\underline{r}) d\underline{r} - \sum_n^{M_i} \int \phi_p^*(\underline{r}) \frac{Z_n}{|\underline{r} - \underline{R}_n|} \phi_q(\underline{r}) d\underline{r} \quad (2.50)$$

$$g_{pqrs} = \iint \phi_p^*(\underline{r}) \phi_r^*(\underline{r}') \frac{1}{|\underline{r} - \underline{r}'|} \phi_q(\underline{r}) \phi_s(\underline{r}') d\underline{r} d\underline{r}'. \quad (2.51)$$

Before elaborating on the interaction operator $\hat{V}^{AB}$, let us define the wave function ansatz for the super-system. Here, we assume the isolated fragment wave functions solve the Schrödinger equation, i.e.,

$$\hat{H}^A |A\rangle = E^A |A\rangle \quad \text{and} \quad \hat{H}^B |B\rangle = E^B |B\rangle, \quad (2.52)$$

and are individually normalized,

$$\langle A|A\rangle = 1 \quad \text{and} \quad \langle B|B\rangle = 1. \quad (2.53)$$

The fragment wave functions can be expressed by a wave operator acting on the vacuum state, that is

$$|A\rangle = \hat{\psi}^A |vac\rangle \quad \text{and} \quad |B\rangle = \hat{\psi}^B |vac\rangle. \quad (2.54)$$

Essentially, the wave operator consists of strings of creation operators, and could define, e.g., Hartree-Fock wave functions. Now we write the supermolecular wave function as a simple product of fragment wave functions,

$$|AB\rangle = \hat{\psi}^A \hat{\psi}^B |vac\rangle, \quad (2.55)$$

by also requiring that the fragment wave operators commute, i.e.,

$$\left[\hat{\psi}^A, \hat{\psi}^B \right] = 0. \quad (2.56)$$

This directly implies that elementary operators acting on different fragments commute, which can be written as

$$[\hat{a}_{p\sigma}^{\dagger A}, \hat{a}_{q\tau}^{\dagger B}] = [\hat{a}_{p\sigma}^{A}, \hat{a}_{q\tau}^{B}] = [\hat{a}_{p\sigma}^{\dagger A}, \hat{a}_{q\tau}^{B}] = 0. \tag{2.57}$$

The last commutator evaluates to zero as it contains vanishing excitation operators which they excite electrons from one fragment to another ($\hat{a}_{p\sigma}^{\dagger A}\hat{a}_{q\tau}^{B} = 0$). However, the anti-commutation relations of the elementary operators still hold fragment-wise.

2.4.2 Interaction Hamiltonian

We can now write the interaction Hamiltonian as

$$\hat{V}^{AB} = \sum_{pq\in A} v_{pq}^{AB} \hat{E}_{pq}^{A} + \sum_{rs\in B} v_{rs}^{AB} \hat{E}_{rs}^{B} + \sum_{\substack{pq\in A \\ rs\in B}} v_{pq,rs}^{AB} \hat{E}_{pq}^{A} \hat{E}_{rs}^{B} + v_{nuc}^{AB},$$

$$\tag{2.58}$$

with inter-fragment nuclear-electron attraction, electron-electron repulsion and nuclear repulsion. Of note, the two-electron excitation operator reduces to a product of two one-electron excitation operators acting on *one* fragment each. The integral v_{pq}^{AB} over the nuclei-electron attraction operator with electrons of fragment A and nuclei of fragment B is

$$v_{pq}^{AB} = - \sum_{m=1}^{M_B} \int \phi_p^{*A}(\mathbf{r}) \frac{Z_m^B}{|\mathbf{r} - \mathbf{R}_m|} \phi_q^A(\mathbf{r}) d\mathbf{r}. \tag{2.59}$$

In the same manner, one finds the complementary integrals for the interaction of electrons of fragment B and nuclei of fragment A as

$$v_{rs}^{AB} = - \sum_{n=1}^{M_A} \int \phi_r^{*B}(\mathbf{r}) \frac{Z_n^A}{|\mathbf{R}_n - \mathbf{r}|} \phi_s^B(\mathbf{r}) d\mathbf{r}. \tag{2.60}$$

The integral over the Coulomb repulsion between electrons in either fragment is given by

$$v_{pq,rs}^{AB} = \iint \phi_p^{*A}(\mathbf{r}) \phi_r^{*B}(\mathbf{r}') \frac{1}{|\mathbf{r} - \mathbf{r}'|} \phi_q^A(\mathbf{r}) \phi_s^B(\mathbf{r}') d\mathbf{r} d\mathbf{r}'. \tag{2.61}$$

Further, we define the nuclear-nuclear repulsion energy as

$$v_{\text{nuc}}^{\text{AB}} = \sum_{n=1}^{M_{\text{A}}} \sum_{m=1}^{M_{\text{B}}} \frac{Z_n^{\text{A}} Z_m^{\text{B}}}{|\mathbf{R}_n - \mathbf{R}_m|}. \tag{2.62}$$

We will now (arbitrarily) focus on fragment A interacting with its environment represented by fragment B. Interaction energy contributions can be derived using standard Rayleigh-Schrödinger perturbation theory on the super-molecular system with the interaction Hamiltonian as the perturbation. Hence, the energies up to second order in the perturbation are given by

$$\mathcal{E}^{(0)} = \langle \text{AB} | \hat{H}^{(0)} | \text{AB} \rangle = E^{\text{A}} + E^{\text{B}} \tag{2.63}$$

$$\mathcal{E}^{(1)} = \langle \text{AB} | \hat{V}^{\text{AB}} | \text{AB} \rangle \tag{2.64}$$

$$\mathcal{E}^{(2)} = - \sum_{\substack{ij \\ i+j \neq 0}} \frac{\langle \text{AB} | \hat{V}^{\text{AB}} | \text{A}^i \text{B}^j \rangle \langle \text{A}^i \text{B}^j | \hat{V}^{\text{AB}} | \text{AB} \rangle}{\epsilon_{ij}^{\text{AB}} - \epsilon_{00}^{\text{AB}}}. \tag{2.65}$$

Obviously, the zeroth-order contribution is just a sum of the unperturbed fragment energies, and the only terms that need further consideration are of higher order. The main problem arising from the given expressions is the explicit dependence on *both* fragment wave functions. To obtain an effective operator, we need to eliminate the wave function of fragment B.

In order to do so, we must first recast the interaction Hamiltonian by a Taylor expansion of the inter-fragment Coulomb interactions around the coordinate $\mathbf{R}_o$ in fragment B,

$$\frac{1}{|\mathbf{r} - \mathbf{r}'|} = \sum_{|k|=0}^{\infty} \frac{(-1)^{|k|}}{k!} \left(\nabla^k \frac{1}{|\mathbf{r} - \mathbf{R}_o|} \right) (\mathbf{r}' - \mathbf{R}_o)^k \tag{2.66}$$

with

$$T_{\text{AB}}^{(k)}(\mathbf{r}) = \nabla^k \frac{1}{|\mathbf{r} - \mathbf{R}_o|}. \tag{2.67}$$

The summation uses a multi-index notation, where k is a 3-tuple, i.e., $k = (k_x, k_y, k_z)$, and the summation runs over $3^{|k|}$ multi-indices.

Exploiting the symmetry of the T-tensors with respect to index permutations (transpositions), there are only $(|k|+1)(|k|+2)/2$ multi-indices. This is also practical for the implementation, however, the multipole moments need to be adapted with the correct pre-factors: For example, the quadrupole tensor Θ would need a factor of 2 on off-diagonal elements, yielding a flat matrix of 6 elements in total: $[\Theta_{xx}, 2\Theta_{yx}, \Theta_{yy}, 2\Theta_{zx}, 2\Theta_{zy}, \Theta_{zz}]$. Removing all redundancies, spherical tensors would only have five matrix elements.

Further, we define $|k| = k_x + k_y + k_z$, $k! = k_x!k_y!k_z!$, and ∇^k is a short-hand notation for the multi-index power of the partial derivative operator

$$\nabla^k = \left(\frac{\partial}{\partial r_x}\right)^{k_x} \left(\frac{\partial}{\partial r_y}\right)^{k_y} \left(\frac{\partial}{\partial r_z}\right)^{k_z} = \frac{\partial^{|k|}}{\partial \mathbf{r}^k} \; . \tag{2.68}$$

$T_{\mathrm{AB}}^{(k)}(\mathbf{r})$ are elements of the so-called *interaction tensor* (also called T-tensor), where (k) specifies the index of multiple tensor elements and the rank of the tensor through $|k|$. Further, the Taylor expansion yields the first-quantization Cartesian multipole moment operator $(\mathbf{r}' - \mathbf{R_o})^k$.

Recasting all Coulomb operators of equation (2.58) in the same manner (see appendix B), a new definition of the interaction Hamiltonian is given by

$$\hat{V}^{\mathrm{AB}} = \sum_{|k|=0}^{\infty} \frac{(-1)^{|k|}}{k!} \left(\mathcal{V}_{\mathrm{A,nuc}}^{(k)} + \hat{\mathcal{V}}_{\mathrm{A,el}}^{(k)}\right) \left(\mathcal{Q}_{\mathrm{B,nuc}}^{(k)} + \hat{\mathcal{Q}}_{\mathrm{B,el}}^{(k)}\right) \tag{2.69}$$

using the operators

$$\mathcal{V}_{A,\text{nuc}}^{(k)} = \sum_{n=1}^{M_A} Z_n^A T_{AB}^{(k)}(\underline{\mathbf{R}}_n) \,, \tag{2.70}$$

$$\hat{\mathcal{V}}_{A,\text{el}}^{(k)} = \sum_{pq \in A} \left(-\int \phi_p^*(\underline{\mathbf{r}}) T_{AB}^{(k)}(\underline{\mathbf{r}}) \phi_q(\underline{\mathbf{r}}) d\underline{\mathbf{r}} \right) \hat{E}_{pq}^A \,, \tag{2.71}$$

$$\mathcal{Q}_{B,\text{nuc}}^{(k)} = \sum_{m=1}^{M_B} Z_m^B \left(\underline{\mathbf{R}}_m - \underline{\mathbf{R}}_o \right)^k \,, \tag{2.72}$$

and

$$\hat{\mathcal{Q}}_{B,\text{el}}^{(k)} = \sum_{rs \in B} \left(-\int \phi_r^*(\underline{\mathbf{r}}') \left(\underline{\mathbf{r}}' - \underline{\mathbf{R}}_o \right)^k \phi_s(\underline{\mathbf{r}}') d\underline{\mathbf{r}}' \right) \hat{E}_{rs}^B \,. \tag{2.73}$$

Plugging in the new definition of the interaction Hamiltonian into equation (2.64), the first-order energy correction is

$$\mathcal{E}^{(1)} = \mathcal{E}_{\text{es}}^{AB} = \sum_{|k|=0}^{\infty} \frac{(-1)^{|k|}}{k!} \left(\mathcal{V}_{A,\text{nuc}}^{(k)} + \langle A | \hat{\mathcal{V}}_{A,\text{el}}^{(k)} | A \rangle \right) \mathcal{Q}_B^{(k)}. \tag{2.74}$$

Here, the expectation value of the multipole moment operator $\mathcal{Q}_B^{(k)}$ was evaluated on fragment B. Thus, the first-order energy describes the plain electrostatic interaction between the fragments, expressed in Cartesian multipole moments of fragment B. From this expression, we can directly obtain the effective PE operator for permanent electrostatics.

The second-order energy needs further analysis: The sum-over-states expression requires at least one system or both systems at once to be in an electronically excited state. Hence, the second-order energy can be split up into three individual contributions as

$$\mathcal{E}^{(2)} = -\sum_{i \neq 0} \frac{\langle AB|\hat{V}^{AB}|A^iB\rangle \langle A^iB|\hat{V}^{AB}|AB\rangle}{\epsilon_i^A - \epsilon_0^A}$$

$$- \sum_{j \neq 0} \frac{\langle AB|\hat{V}^{AB}|AB^j\rangle \langle AB^j|\hat{V}^{AB}|AB\rangle}{\epsilon_j^B - \epsilon_0^B} + \mathcal{E}_{disp} \qquad (2.75)$$

$$= \mathcal{E}_{ind}^A + \mathcal{E}_{ind}^B + \mathcal{E}_{disp} . \qquad (2.76)$$

The induction energy of fragment A, $\mathcal{E}_{ind}^A$, need not further be taken into account since it is implicitly included through the effective PE operator in the wave function optimization (orbital rotations) such that $|A\rangle$ is being polarized by the multipole moments of fragment B. Induction energy is a classical energy term. It is contained in the electronic density of the molecular system through orbital rotations that build the electronic ground state. There is no straightforward way to define an induction energy quantum-mechanically or to extract this energy from the description of the electronic ground state. The last term in equation (2.76), $\mathcal{E}_{disp}$, denotes the dispersion energy (both systems are in an excited state). Since there is no well-defined way to include this energy contribution in operator form, it is neglected. Note that an empirical dispersion correction using a Lennard-Jones potential between fragment A and B can be introduced in an *ad-hoc* manner.

Finally, we can evaluate the expectation value of the interaction Hamiltonian of the second term in equation (2.76), namely when fragment B is in an excited state and fragment A is in the electronic ground state. This yields

$$
\mathcal{E}_{\text{ind}}^{\text{B}} = - \sum_{|k|=0}^{\infty} \frac{(-1)^{|k|}}{k!} \langle A | \hat{\mathcal{V}}_{\text{A}}^{(k)} | A \rangle
$$

$$
\sum_{j \neq 0} \frac{\langle B | \hat{\mathcal{Q}}_{\text{B}}^{(k)} | B^j \rangle \, \langle B^j | \hat{\mathcal{Q}}_{\text{B}}^{(k)} | B \rangle}{\epsilon_j^{\text{B}} - \epsilon_0^{\text{B}}} \langle A | \hat{\mathcal{V}}_{\text{A}}^{(k)} | A \rangle . \qquad (2.77)
$$

The sum-over-states expression describes the polarizabilities of fragment B. The zeroth-order term does not contribute due to orthogonality of the states, i.e., $\langle B | q_{\text{B}} | B^j \rangle = 0$. When truncating the expansion at the dipole level, i.e., $|k| = 1$, and evaluating the respective expectation values on fragment A, we get

$$
\mathcal{E}_{\text{ind}}^{\text{B}} = -\frac{1}{2} \left(\underline{\mathbf{F}}_{\text{nuc}}^{\text{A}} + \underline{\mathbf{F}}_{\text{el}}^{\text{A}} \right)^{\text{T}} \boldsymbol{\alpha}^{\text{B}} \left(\underline{\mathbf{F}}_{\text{nuc}}^{\text{A}} + \underline{\mathbf{F}}_{\text{el}}^{\text{A}} \right)
$$

$$
= -\frac{1}{2} \left(\underline{\mathbf{F}}_{\text{nuc}}^{\text{A}} + \underline{\mathbf{F}}_{\text{el}}^{\text{A}} \right) \underline{\boldsymbol{\mu}}_{\text{ind}}^{\text{B}} (\underline{\mathbf{F}}_{\text{tot}}). \qquad (2.78)
$$

In this context, the total electric field stems from the electrons ($\underline{\mathbf{F}}_{\text{el}}^{\text{A}}$) and nuclei ($\underline{\mathbf{F}}_{\text{nuc}}^{\text{A}}$) of the quantum region, however, the expression can be extended for multiple expansion sites in the environment. Of note, the induction energy is non-additive since it explicitly depends on the electron density of fragment A in a non-linear manner.

2.4.3 The Effective PE Operator

Finally, one is able to derive an effective PE operator by generalizing the equations for S multipole expansion sites in the environment.

The electrostatics operator is already 'well prepared' and can be easily extended for an arbitrary number of sites, whereas the induction operator is obtained from electric fields created by induced moments. Determination of the induced dipole moments is a little more involved: The induced dipole moment at a given site depends on *all* other induced dipole moments.

The total PE energy functional[15,16] is given by

$$
\mathcal{E}_{\text{tot}} = \mathcal{E}_{\text{QM}} + \mathcal{E}_{\text{PE}} + \mathcal{E}_{\text{env}}, \qquad (2.79)
$$

where $\mathcal{E}_{QM}$ is the energy of the quantum region, including polarization of the wave function. $\mathcal{E}_{PE}$ represents the interaction energy of the quantum region and the environment, including polarization of the environment. $\mathcal{E}_{env}$ denotes the internal energy of all fragments in the environment, including interaction among fragments but excluding energy contributions from polarization in the environment. Decomposing the polarizable embedding energy, one finds

$$\mathcal{E}_{PE} = \mathcal{E}_{es} + \mathcal{E}_{ind}, \qquad (2.80)$$

where $\mathcal{E}_{es}$ denotes the permanent electrostatic interaction between the core quantum part and the fragments in the environment, and $\mathcal{E}_{ind}$ is the energy due to induced charge distributions of the environment. The explicit equation for the electrostatic interaction energy is

$$\mathcal{E}_{es} = \mathcal{E}_{es}^{nuc} + \mathcal{E}_{es}^{el}, \qquad (2.81)$$

comprised by the nuclear and electronic interaction energies, respectively, where the former is given by

$$\mathcal{E}_{es}^{nuc} = \sum_{s=1}^{S} \sum_{|k|=0}^{K_s} \frac{(-1)^{|k|}}{k!} Q_s^{(k)} \sum_{n=1}^{N} T_{sn}^{(k)} Z_n. \qquad (2.82)$$

The summation over $|k|$ is running over $3^{|k|}$ multi-indices up to the truncation level K_s of the multipole expansion and the summation over s is running over the S sites in the environment. The $Q_s^{(k)}$ is thus the k-th component of the $|k|$-th-order Cartesian multipole moment located at the expansion site coordinate $\mathbf{R}_s$ in the environment, and Z_n is the nuclear charge of the n-th nucleus in the quantum region comprised of N nuclei. As before, we have used the k-th component of the interaction tensor, $T_{ij}^{(k)}$, between two sites i and j,

$$T_{ij}^{(k)} = \frac{\partial^{|k|}}{\partial x_j^{k_x} \partial y_j^{k_y} \partial z_j^{k_z}} \left(\frac{1}{|\mathbf{r}_j - \mathbf{r}_i|} \right). \qquad (2.83)$$

Further, the electrostatic interaction energy of the electrons with the environment is given by the expectation value of the electrostatic operator $\hat{\mathcal{V}}_{es}$ using the ground state determinant $|0\rangle$, that is

$$\mathcal{E}_{es}^{el} = \langle 0|\hat{\mathcal{V}}_{es}|0\rangle . \tag{2.84}$$

Using the second-quantization formalism, we can write the electrostatic operator, $\hat{\mathcal{V}}_{es}$, as

$$\hat{\mathcal{V}}_{es} = \sum_{s=1}^{S} \sum_{|k|=0}^{K_s} \frac{(-1)^{|k|}}{k!} Q_s^{(k)} \sum_{pq} t_{pq}^{(k)}(\underline{\mathbf{R}}_s)\hat{E}_{pq}, \tag{2.85}$$

with the one-electron orbital excitation operator $\hat{E}_{pq}$ and general molecular orbital indices p and q. The integrals are given by

$$t_{pq}^{(k)}(\underline{\mathbf{R}}_s) = -\int \phi_p^*(\underline{\mathbf{r}}_1)T_{s1}^{(k)}\phi_q(\underline{\mathbf{r}}_1)d\underline{\mathbf{r}}_1, \tag{2.86}$$

and include again the k-th component of the interaction tensor (eq (2.83)).

The induction energy contribution of a linearly responsive environment amounts to

$$\mathcal{E}_{ind} = -\frac{1}{2}\sum_{s=1}^{S}\underline{\mu}_s^{ind}(\underline{\mathbf{F}})^{\mathrm{T}}\mathbf{F}(\underline{\mathbf{R}}_s), \tag{2.87}$$

where $\underline{\mu}_s^{ind}(\underline{\mathbf{F}})$ is the induced dipole moment at site s in the environment, and $\underline{\mathbf{F}}(\underline{\mathbf{R}}_s)$ is the electric field vector acting on site s, comprising the field from nuclei and electrons, as well as the fields caused by the permanent multipole moments, i.e.,

$$\underline{\mathbf{F}}[\mathbf{P}] = \underline{\mathbf{F}}_{nuc} + \underline{\mathbf{F}}_{el}[\mathbf{P}] + \underline{\mathbf{F}}_{mul} . \tag{2.88}$$

Note that the electric field from the electrons, and in turn the total field vector, $\underline{\mathbf{F}}$, *explicitly* depend on the electronic density matrix $\mathbf{P}$.

We find that the induced moment at a site s depends on the total electric field and is given by

$$\boldsymbol{\mu}_s^{\text{ind}}(\underline{\mathbf{F}}_{\text{tot}}) = \alpha_s \left(\underline{\mathbf{F}}_s[\mathbf{P}] + \underline{\mathbf{F}}_s^{\text{ind}} \right). \tag{2.89}$$

Here, we have added the induced fields created by all other sites, i.e.,

$$\underline{\mathbf{F}}_s^{\text{ind}} = \sum_{s' \neq s} \mathbf{T}_{ss'}^{(2)} \boldsymbol{\mu}_{s'}^{\text{ind}}(\underline{\mathbf{F}}_{\text{tot}}). \tag{2.90}$$

This leads to a linear system of equations,

$$\mathbf{B}\underline{\boldsymbol{\mu}}^{\text{ind}}(\underline{\mathbf{F}}_{\text{tot}}) = \underline{\mathbf{F}}[\mathbf{P}], \tag{2.91}$$

with the classical response matrix $\mathbf{B}$[53] (also called relay matrix), given by

$$\mathbf{B} = \begin{pmatrix} \boldsymbol{\alpha}_1^{-1} & -\mathbf{T}_{12}^{(2)} & \cdots & & -\mathbf{T}_{1S}^{(2)} \\ -\mathbf{T}_{21}^{(2)} & \boldsymbol{\alpha}_2^{-1} & \ddots & & \vdots \\ \vdots & \ddots & \ddots & & -\mathbf{T}_{(S-1)S}^{(2)} \\ -\mathbf{T}_{S1}^{(2)} & \cdots & & -\mathbf{T}_{S(S-1)}^{(2)} & \boldsymbol{\alpha}_S^{-1} \end{pmatrix}. \tag{2.92}$$

The inverse site polarizability tensors $\boldsymbol{\alpha}_s^{-1}$ are on the diagonal and the dipole-dipole interaction tensors reside on off-diagonal elements. Subsequently, we can include the induced dipole field into the wave function optimization through the induction operator

$$\hat{\mathcal{V}}_{\text{ind}}[\mathbf{P}] = -\sum_{s=1}^{S} \sum_{a=x,y,z} \mu_{a,s}^{\text{ind}}(\underline{\mathbf{F}}_{\text{tot}}) \hat{F}_a^e(\underline{\mathbf{R}}_s), \tag{2.93}$$

using a for the respective Cartesian component $x, y,$ or z. Further, we define the electric-field operator as

$$\hat{F}_a^e(\underline{\mathbf{R}}_s) = \sum_{pq} t_{a,pq}(\underline{\mathbf{R}}_s) \hat{E}_{pq}. \tag{2.94}$$

The electric-field integrals are defined as

$$t_{a,pq}(\underline{\mathbf{R}}_s) = -\int \phi_p^*(\underline{\mathbf{r}}) \frac{R_{a,s} - r_a}{|\underline{\mathbf{R}}_s - \underline{\mathbf{r}}|^3} \phi_q(\underline{\mathbf{r}}) d\underline{\mathbf{r}}. \qquad (2.95)$$

Finally, one solves the self-consistent field (SCF) problem in the presence of the total embedding operator,

$$\hat{\mathcal{V}}_{\mathrm{PE}}[\mathbf{P}] = \hat{\mathcal{V}}_{\mathrm{es}} + \hat{\mathcal{V}}_{\mathrm{ind}}[\mathbf{P}], \qquad (2.96)$$

from which we obtain an effective Fock operator

$$\hat{\mathcal{V}}_{\mathrm{HF,eff}} = \hat{\mathcal{V}}_{\mathrm{HF}} + \hat{\mathcal{V}}_{\mathrm{PE}}. \qquad (2.97)$$

Since the embedding operator depends on the wave function itself, namely through the electric fields, the overall embedding operator is non-linear. As for a usual SCF procedure, where the exchange and Coulomb terms depend on the density, the embedding operator is updated in every iteration using the current SCF density matrix. Thus, polarization effects are treated in a fully self-consistent manner for the electronic ground state.

3 Combination of PE and ADC

My approach to include contributions from PE in ADC calculations of excitation energies is fully density-driven and based on perturbative, *a posteriori* corrections. As a consequence, the expressions for the ADC secular matrix (eq (2.43)) elements do not change. A similar approach was pursued for the combination of PCM with ADC[40,41] or the effective fragment potential (EFP)[54–58] method with EOM-CCSD[59,60], for example. In recent publications[17,42,61], it has been argued that both state-specific and linear-response-type contributions should be accounted for in order to obtain accurate excitation energies. I will follow these considerations and in the following assume the environment to remain in the electronic ground state, thus, the electronic excitation to be localized on the core molecule. Generally, the shift in excitation energy of the core molecule in presence of an environment, also called *perichromatic* shift, can be written up to second order[61] as

$$\Delta E_{\text{shift}}^{0\to n} = \underbrace{\Delta E_{\text{es}}^{0\to n}}_{\text{(i)}} + \underbrace{\Delta E_{\text{ind}}^{0\to n}}_{\text{(ii)}} + \underbrace{\Delta E_{\text{disp}}^{0\to n}}_{\text{(iii)}} + \underbrace{\Delta E_{\text{excoupl}}^{0\to n}}_{\text{(iv)}}. \qquad (3.1)$$

The individual terms denote (i) the difference in Coulomb interaction energy with the electrostatic environment, (ii) the change in mutual induction energy upon excitation, (iii) energy difference in London dispersion energy, and (iv) the non-resonant excitonic coupling. In the PE model, London dispersion energy is not introduced by construction, so term (iii) is neglected. Moreover, these contributions have been argued to be rather small[17,61]. Since the PE model includes electrostatics, term (i) is already accounted for in the reference HF ground state entering the ADC calculation. This leaves us with the

© The Editor(s) (if applicable) and The Author(s), under exclusive license to
Springer Fachmedien Wiesbaden GmbH, part of Springer Nature 2020
M. Scheurer, *Polarizable Embedding for the Algebraic- Diagrammatic
Construction Scheme*, BestMasters, https://doi.org/10.1007/978-3-658-31281-7_3

change in induction energy (ii) and non-resonant excitonic coupling (iv). The change in induction energy depends on the density matrix of the excited state and the resulting induced moments in the environment. The induced moments enter the ADC problem as a static dipole field through the Fock operator since the electric field of the electrons on environmental sites is not being updated during the ADC procedure. Hence, the change in induction energy included by construction amounts to. Products of two vectors are implicitly scalar products in the following equations. The notation omits a transposition on the left vector for better readability.

$$\Delta E(0)_{\text{el,ind}}^{0\to n} = -\Delta \underline{\mathbf{F}}_{\text{ind}}^{0\to n} \underline{\boldsymbol{\mu}}_{\text{ind}}^0 \tag{3.2}$$

$$= -\underline{\mathbf{F}}_{\text{el}}^i \underline{\boldsymbol{\mu}}_{\text{el,ind}}^0 + \underline{\mathbf{F}}_{\text{el}}^0 \underline{\boldsymbol{\mu}}_{\text{el,ind}}^0, \tag{3.3}$$

with $\Delta E(0)_{\text{el,ind}}^{0\to n}$ emphasizing the *frozen* induced moments of the ground state. The perichromatic shift with just these terms, i.e., a frozen PE ground state, is expressed by

$$\Delta E(0)_{\text{shift}}^{0\to n} = \Delta E_{\text{es}}^{0\to n} + \Delta E(0)_{\text{el,ind}}^{0\to n} \tag{3.4}$$

with only zeroth-order contributions to the excitation energies. To improve upon this description, induced moments created through the density of the excited state need to be taken into account, that is, a correction term will be added to eq (3.3) using a state-specific approach[61]. The total induction energy through the electric fields $\underline{\mathbf{F}}_{\text{el}}^n$ in a given state n due to electrons (nuclear and multipole fields do not change upon excitation) is given by

$$E_{\text{ind}}^n = -\frac{1}{2}\underline{\mathbf{F}}_{\text{el}}^n \underline{\boldsymbol{\mu}}_{\text{el,ind}}^n, \tag{3.5}$$

and the difference in induction energy between the ground state and an excited state is simply

$$\Delta E_{\text{ind}}^{0\to n} = -\frac{1}{2}\left(\underline{\mathbf{F}}_{\text{el}}^n \underline{\boldsymbol{\mu}}_{\text{el,ind}}^n - \underline{\mathbf{F}}_{\text{el}}^0 \underline{\boldsymbol{\mu}}_{\text{el,ind}}^0\right). \tag{3.6}$$

Taking the difference of eq (3.3) and the above equation, one finds the perturbative state-specific (ptSS) correction term for the excitation energy as

$$\Delta E_{\text{ptSS}}^{0 \to n} = -\frac{1}{2} \underline{F}_{\text{el}}[\Delta \mathbf{P}] \underline{\mu}_{\text{ind}}[\Delta \mathbf{P}]. \qquad (3.7)$$

For a comprehensive derivation, see Ref. 17. The last term in eq (3.1) missing in the description is the non-resonant excitonic coupling energy contribution (iv). As has been shown in great detail using long-range perturbation theory by Schwabe[61], and described in a similar manner for PCM[62], the non-resonant excitonic coupling describes the interaction of the transition dipole moment with the induced moments in the environment, consistently taken into account through response theory approaches[15–17,61]. Since we do not include the embedding operator explicitly in the ADC problem, we correct for non-resonant excitonic coupling with a perturbative linear-response-type (ptLR) expression,

$$\Delta E_{\text{ptLR}}^{0 \to n} = -\underline{F}_{\text{el}}[\mathbf{P}^{\text{tr}}] \underline{\mu}_{\text{ind}}[\mathbf{P}^{\text{tr}}]. \qquad (3.8)$$

Here, couplings between different excited states are neglected, as done in preceding work as well[17,61]. In total, the correction for the n-th excitation energy amounts to

$$\Delta E_{\text{corr}}^{0 \to n} = \Delta E_{\text{ptSS}}^{0 \to n} + \Delta E_{\text{ptLR}}^{0 \to n}. \qquad (3.9)$$

To summarize, contributions from the environment using the PE approach are included indirectly through the frozen HF ground state and excitation energies are corrected *a posteriori* with perturbative treatments (eq (3.9)) using the difference and transition densities of the obtained excited states. All excited state properties are, however, calculated without further corrections in presence of the frozen PE-HF ground state.

4 Implementation

With the theory underlying a possible combination of PE and ADC
being well established, the PE model must be implemented in the
Q-Chem program package to obtain a PE-HF ground state and ex-
ecute a subsequent ADC calculation with adcman. The original PE
implementation resides in PElib, a FORTRAN library used by *Dalton*[63].
For this thesis, I have taken the approach to implement a novel, stand-
alone library, named CPPE, providing the necessary toolkit to carry
out PE calculations in any quantum chemical host program. This
novel library was interfaced with the *Q-Chem* program package to
enable PE-ADC calculations. Until now, CPPE has been interfaced to
four quantum-chemical program packages, namely *Q-Chem*, Psi4[64],
PySCF[65], and VeloxChem[66]. The library features, design, and in-
terfaces to other program packages are presented in great detail in my
publication[67]. In the course of this thesis, the design goals presented
in the following have emerged. They have since been overhauled and
amended[67], e.g., with an easy-to-use python interface. Nevertheless,
the following, original design goals were key to the initial design of
CPPE. The first version of the CPPE library is described afterwards, for
more up-to-date details, I refer the reader to Ref. 67.

4.1 Motivation and Design Goals

As mentioned above, the CPPE library should provide the necessary
functionality and building blocks to incorporate the PE model in a
multitude of program packages. Modern quantum chemistry program
packages are mostly written in the C++ programming language (e.g.,
Q-Chem[68] and *ORCA*[69]), whereas more traditional packages still use

© The Editor(s) (if applicable) and The Author(s), under exclusive license to
Springer Fachmedien Wiesbaden GmbH, part of Springer Nature 2020
M. Scheurer, *Polarizable Embedding for the Algebraic- Diagrammatic
Construction Scheme*, BestMasters, https://doi.org/10.1007/978-3-658-31281-7_4

FORTRAN. The more recent trend goes toward usage of python[64,65,70] as a high-level, easy-to-use scripting language for rapid prototyping.

For FORTRAN-only programs, the already available PElib is in principle sufficient to run PE calculations as the required integrals are evaluated from the gen1int library. In principle, PElib should be independent of the host program, however, it uses *Dalton* routines to call the gen1int integral engine, which makes it almost impossible to use the library in a plug-and-play manner in another code base. Further, PElib is parallelized using MPI, whereas a multitude of quantum chemistry programs rely on shared-memory OMP parallelization. Furthermore, the molecular integral calculation with gen1int is deeply buried inside PElib, i.e., the integrals cannot easily be provided by the host program. This would be of advantage, e.g., if the kind of parallelization between PElib and the host program do not match. In regard of a possible python interface for PElib, the design of the library prohibits easy serialization with cython or SWIG[71,72]. Given these pitfalls and shortcomings of existing PE implementations, the design goals of the CPPE library can be elaborated:

a.) *Plug-and-play interface to* PElib
As a matter of fact, PElib not only implements standard ground-state PE calculations, but also the necessary toolkit to obtain molecular gradients, magnetic gradients and environment contributions needed to obtain molecular response properties. Further, it implements the polarizable density embedding (PDE) method, as well as the fluctuating charges (FQ) model[73]. Also, creation of *cube* files for subsequent density visualization is available. It is hence beneficial for any host program to interface with PElib. This requires removal of *Dalton* dependencies for calls to the integral routines in gen1int in combination with an interface to PElib for non-FORTRAN host programs. Such an interface would make all future features of PElib directly available without changing much code inside the host program. Thus, the goal is to make PElib usable in a plug-and-play manner to a plethora of host programs.

b.) *Flexible, host-program-independent high-level interface*
In the context of novel method design and rapid prototyping, it is required to have an interface which is not only accessible through low-level programming languages, such as C++, but also through high-level scripting languages as python. A plethora of python-based quantum chemistry programs have emerged over the last years,[64,65] paving the way for rapid method development, new combined approaches, and education in quantum chemistry. Since embedding models, such as PE, can in principle be combined with any quantum chemical method, quick testing of novel combinations should be feasible using the new library. However, this requires straightforward serialization of data structures between the high-level and low-level code. For performance reasons, the computationally demanding routines for PE as well as the molecular integral codes should be written in a compiled language, such as C++. This directly leads to the next design goal.

c.) *Molecular integrals from an arbitrary integral library*
Integral engines are key to the performance of quantum chemical program package. Many packages ship with their own integral engines, highly optimized for the inherent structure of the packages as such, as well as the underlying parallelization paradigm. Even though the existing PElib is directly interfaced with genlint, parallelization is only achieved using distributed-memory MPI. Hence, no shared-memory-parallel implementation is available, even though many program packages, such as *Q-Chem*, mainly operate in an OMP-parallel manner. For this reason, compatibility with respect to parallelization of integral calculations is not ideal. It should therefore be feasible to feed the CPPE library integrals from the host program to exploit performance of the respective integral engine, no matter what parallelization paradigm is used. This grants enormous flexibility also in regard of the host-program-independent high-level interface.

d.) *Low code complexity*
Last but not least, the code should be of low complexity for various

reasons. First, the interface to PElib should be easily extensible to support novel features in PElib itself. Second, the CPPE library must provide building blocks for easy experimentation with the existing PE model as well as related and future variants of PE. This is not possible in PElib due to the lack of modern programming paradigms, e.g., object-oriented programming, in FORTRAN. Moreover, the library should allow for intuitive manipulation of input embedding potentials. The existing PElib code for this task is hard to read or extend, in my opinion, and an object-oriented programming language would allow for much easier, well-structured, and readable code. Also, serialization to high-level scripting languages would be facilitated if a language like C++ were used.

4.2 Pilot Version of the CPPE Library

The first version of the CPPE library was implemented based on the above design goals using C++ as the main programming language. This enabled an interface to the low-level FORTRAN code of PElib, but also incorporates a stand-alone C++ implementation of PE routines exploiting the benefits of object-oriented programming and polymorphism. Furthermore, the C++ implementation of needed PE routines could be easily linked to python through pybind11. This approach unifies the design goals, and it results in two possible modes of carrying out PE calculations: The first mode comprises running PElib in a 'black-box' manner using integrals from gen1int. On the contrary, the second mode requires input of the needed molecular integrals from the host program and uses the new C++ implementation of PE routines. In the first run mode, CPPE acts as a wrapper for PElib by processing and transferring all necessary data from the host program to the low-level FORTRAN routines. In the second mode, however, CPPE does not make any use of PElib at all, but carries out the necessary file read-in and calculations in a stand-alone manner. The main building blocks of CPPE, including the top-level interface for the host program, the FORTRAN-interface to PElib, and the new C++-based implementation of PE routines are described in the following. First, I will briefly introduce the overall structure of CPPE and how it is intended to be used in a quantum chemical program package. Subsequently, the lower-level implementations are described in more detail. Figure 4.1

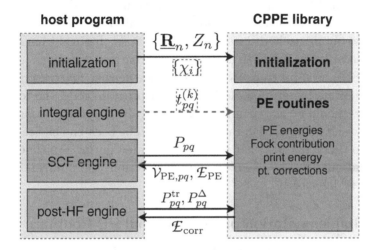

Figure 4.1: CPPE interface overview. In the red box on the left, the key host program components are listed. On the right, the box corresponding to CPPE is shown. Arrows with formula symbols depict the data that is being exchanged between the host program and CPPE. Black arrows are common for both back-ends, whereas dashed pink lines indicate data that is exclusive for one of the back-ends.

depicts a general overview of the PE calculation workflow using CPPE. Upon host program initialization, the user input is being parsed and options for PE, including nuclear coordinates ($\{\underline{\mathbf{R}}_n\}$) and charges ($\{Z_n\}$) are passed to CPPE. In case the PElib back-end is used, the host program must also provide the AO basis ($\{\chi_i\}$, dashed pink box) for genlint initialization. If the CPPE-internal back-end is used, the molecular integrals ($t_{pq}^{(k)}$, eq (2.86)) over the Coulombic operator must be provided by the host program's integral library (or another external integral library). To obtain the ground state wave function (HF or

DFT) in presence of the polarizable environment, the PE operator needs to be included in the wave function optimization, i.e., the SCF engine. The CPPE needs the density matrix (P_{pq}) of the current SCF iteration in the AO basis to build the effective PE operator $(\hat{\mathcal{V}}_{\text{PE},pq})$ and calculate the respective energy contributions $(\mathcal{E}_{\text{PE}})$. Once the SCF has reached convergence, post-HF calculations such as ADC can be run to obtain excited states using the orbitals obtained from PE-HF. CPPE can again be called to calculate perturbative corrections (eq (3.7) and (3.8)) based on transition or difference density matrices.

To facilitate integration of CPPE in host program code, the top-level interface should be as general as possible to abstract the interface from the back-end being used in the calculation. The back-end choice should be up to the user, such that it must be decided during run time which back-end ought to be used. In the next section, the implementation thereof is described.

4.2.1 Implementation of CPPE-internal PE Routines

To comply with the presented design goals, a set of C++ classes providing the necessary building blocks and math functionality was implemented. Here, we can distinguish between `utils` and `core` functionality. I have decided to use `armadillo`[74] as linear algebra library since it is template-based and allows intuitive handling of matrices and vectors, also for complex algebra. The `utils` classes, such as `PotfileReader` and `PotManipulator` are capable of parsing potential files and manipulating the embedding potential, respectively. Upon read-in, the potential file is stored in a `std::vector<Potential>`, i.e., a vector of `Potentials`. The `Potential` class is illustrated in Figure 4.2. Essentially, an instance of `Potential` holds all necessary information of *one* specific site in the environment, including its multipole moments, polarizabilities, exclusion list, and coordinate. The class is easily extensible and allows for straightforward usage of the potential in down-stream methods. Its class methods provide the necessary toolkit (`bool is_polarizable()`, `bool excludes_site(int other_site)`) for a light-weight implementation of lower-level methods, e.g., calculation of electrostatic interaction energies. All this is in contrast to the FORTRAN implementation in PElib, where the data is only accessible through order-specific, hard-coded arrays of multipole moments or polarizabilities. Hence, the Potential class in CPPE provides a much more intuitive and generalized way of dealing with the molecular environment.

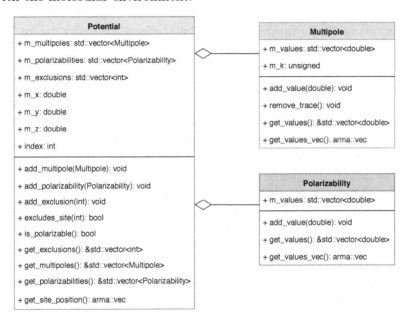

Figure 4.2: Design of the Potential, Multipole, and Polarizability classes. The fields of Potential hold lists of multipole moments and dipole-dipole polarizabilities. Further, the coordinates m_x, m_y, and m_z of the site are stored, together with the site index. The first three instance methods (starting with add_) are used by the PotfileReader. Multipole implements a container for values of a multipole moment of a certain order m_k. Polarizability implements a dipole-dipole polarizability.

Another cornerstone of the CPPE core implementation is the CppeState: It acts both as a container of all necessary meta-data (operators, energies, field vectors) and a wrapper for the necessary core functionality. This class facilitates implementation of a `PeCalcHandler` subclass for usage of CPPE-internal PE routines. It is able to receive the integrals in matrix representation from the host program and carry out all subsequent steps provided these integrals (see design goals *b)* and *c)*). An instance of `CppeState` is created when the `CppePeCalcHandler` is initialized and provided the core molecule and embedding potential. Thus, all methods inside the `CppePeCalcHandler` make use of the `CppeState` instance. For example, if the static energies and fields are to be calculated, one can do so by simply calling `CppeState::calculate_static_energies_and_fields()`.

An additional key component in `CppeState` is the `update_induced_moments(...)` method. It assembles the total field vector, depending on which electric fields should be taken into account for the induced moments, and afterwards sets up the `InducedMoments` solver. After the solver has been run, the polarization energies are updated in the `PeEnergy` object, which is also a field of `CppeState`. `PeEnergy` is the container to hold all different PE energy contributions which are accessible through simple strings. For example, to get the polarization energy through the nuclear fields, one simply calls `m_pe_energy.get("Polarization/Nuclear");`. This makes the code more readable and facilitates print-out of the embedding energies.

Solver for Induced Moments

The linear system of equations to determine the induced moments is solved with the Gauss-Seidel method iterating over the equation

$$\underline{\mu}_s^{[i]} = \alpha_s \left(\underline{\mathbf{F}}(\underline{\mathbf{R}}_s) + \sum_{s'<s} \mathbf{T}_{s's}^{(2)} \underline{\mu}_{s'}^{[i]} + \sum_{s'>s} \mathbf{T}_{s's}^{(2)} \underline{\mu}_{s'}^{[i-1]} \right). \qquad (4.1)$$

Here, $[i]$ denotes the current iteration, such that $\underline{\mu}^{[i-1]}$ is an induced moment from a previous iteration. The implementation of the solver resides in

`InducedMoments::compute(...)`. To start the solver, a guess vector for the induced moments is created as the scalar product of the site polarizability and the static field, i.e.,

$$\underline{\mu}_s^{[0]} = \alpha_s \underline{\mathbf{F}}(\underline{\mathbf{R}}_s).$$

After convergence, the induced moments vector is stored in memory and used as a guess in the following SCF iteration. The convergence is further improved through Anderson mixing (often called DIIS acceleration). To implement different types of linear solvers, a class representing the **B** matrix was implemented in CPPE, providing a straightforward matrix-vector apply.

4.3 Code Availability

To summarize, the above implementation of CPPE fulfills the design goals and presents a flexible, modern, and light-weight implementation of PE routines. As of today, the library is used in at least four program packages[67], which has greatly improved the availability of the PE model in open-source codes. The code of the CPPE library is freely available for download on `https://github.com/maxscheurer/cppe` under the LGPLv3 license.

5 Computational Methodology

The implementation described in the previous Chapter was interfaced with a development version of the *Q-Chem* program package[68] based on version 5.1. In this Chapter, the computational methodology of the first PE-ADC benchmark calculations and applications are thoroughly described. Data analysis for all calculations was performed using the Python packages cclib[75], orbkit[76], pandas[77,78], numpy[79], scipy[80]. Matplotlib[81] and seaborn[82] were used to create plots.

5.1 Para-Nitroaniline with Water Clusters

Para-Nitroaniline (PNA) structures with two, four and six water molecules were used from previous work by Slipchenko[59]. The three lowest singlet excitation energies were calculated at the ADC(2) and ADC(3) levels of theory using Dunning's cc-pVDZ basis set[83–85] for the super-system including solvent, PNA in vacuum, and PE-ADC(2) with ptSS and ptLR corrections. PyFrame[86] was used to create embedding potentials for the solvent water molecules based on LoProp[87] multipole moments through second order and anisotropic dipole-dipole polarizabilities. LoProp properties were calculated at the PBE0/cc-pVDZ level of theory[88] using the Dalton program[63] and the LoProp for Dalton script[89]. Detachment and attachment densities for the lowest $n\pi^*$ (1^1A_2) and $\pi\pi^*$ (1^1A_1) excitations were obtained using the libwfa library[90,91] and rendered with VMD[92].

© The Editor(s) (if applicable) and The Author(s), under exclusive license to
Springer Fachmedien Wiesbaden GmbH, part of Springer Nature 2020
M. Scheurer, *Polarizable Embedding for the Algebraic- Diagrammatic*
Construction Scheme, BestMasters, https://doi.org/10.1007/978-3-658-31281-7_5

5.2 Lumiflavin in Bulk Solvent

To compare the performance of a continuum solvation model linked to ADC(2) with the PE-ADC approach, a benchmark study was performed on the flavin derivative lumiflavin (Lf). Excitation energies at the ADC(2)/cc-pVDZ level of theory were computed using three different setups, summarized in Table 5.1. Therein, water (H_2O) was selected as a polar, protic solvent, and cyclohexane (cyHex) as an apolar, aprotic solvent. The three-parameter corrected HF (HF-

Table 5.1: Benchmark systems for Lf in bulk solvent.

#	geometries	excited states	solvent
1	HF-3c	ADC(2)	none
2	HF-3c/PCM	PCM-ADC(2)	H_2O, cyHex
3	QM/MM MD	PE-ADC(2)	H_2O, cyHex

3c) method[93] as implemented in the Orca 4 program package[69,94] was used to optimized the first two systems, whereas QM/MM MD was used for sampling with the same method to obtain the input geometries for PE-ADC. A general workflow of how to simulate UV/Vis spectra with PCM-ADC or PE-ADC is depicted in Figure 1. Note that explicit solvent modeling, i.e., PE-ADC, requires configurational sampling of the solvent molecules through MD and/or QM/MM MD simulations. By contrast, implicit solvent modeling with PCM-ADC usually employs quantum mechanically optimized structures since the sampling is implicitly contained in the model by virtue of the macroscopic dielectric constants. This is, of course, not mandatory, and PCM calculations could in general be performed on any molecular geometry. Of note, the quantum-chemical method used for sampling in QM/MM MD simulations must be carefully chosen to provide suitable input geometries for high-level *ab-initio* excited states calculations. It is hence possible that the order, energy and character of excited states is affected by the method used for configurational sampling. The most reliable and consistent approach would be to use the underlying ground state method for configurational sampling. This is, by all means, prohibitive for highly accurate excited states methods since the nuclear gradient calculations for correlated methods are usually time- and memory-consuming. For ADC(2), e.g., one would have to run the QM/MM MD with MP(2) gradients.

Table 5.2: Comparison of MP2 and HF-3c geometries in ADC(2) calculations for Lf.

excitation	MP2/cc-pvDZ[a]		HF-3c[b]	
	E [eV]	f	E [eV]	f
1	2.923	0.259	3.230	0.257
2	3.354	0.000	3.610	0.001
3	3.589	0.001	3.804	0.000
4	4.144	0.166	4.498	0.121

Mean blue shift: 0.283 eV

[a] MP2 as implemented in the Q-Chem program package, version 5
[b] HF-3c as implemented in the Orca program package, version 4

Note that HF-3c is being used here for fast sampling in the QM/MM MD simulations and that excitation energies compared to MP(2)-optimized structures are blue-shifted by approximately 0.3 eV (Tab. 5.2). The detailed protocol for MD and QM/MM MD configurational sampling is thoroughly explained in the following.

QM/MM MD setup: Lf was placed in a cube of 50 Å edge length filled with water molecules or cyHex using Packmol[95]. For configurational sampling, MD simulations of these systems were carried out using NAMD 2.12[96]. The CHARMM36 force field[97] was used for solvents and Lf was parametrized using SwissParam[98]. System setup was performed with the *psfgen* plugin in VMD[92]. The systems were first minimized for 10000 steps and then simulated in the NpT

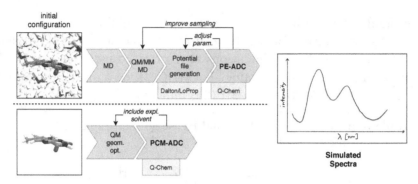

Figure 5.1: Workflows for PE-ADC and PCM-ADC calculations. From
an initial chromophore configuration with explicit solvent,
the workflow towards PE-ADC is described in the upper
diagram, whereas implicit solvent modeling is described in
the lower diagram. The PE-ADC method requires configurational sampling through MD and QM/MM MD simulations
and subsequent parametrization of the environment. Results
can be improved through longer sampling and adjusting the
level of parametrization. Initial structures ought to be optimized first before entering a PCM-ADC calculation, and
spectra can be refined by adding explicit solvent molecules to
the QM system.[1]

ensemble with periodic boundary conditions for 5 ns using an integration time step of 2 fs. Temperature was maintained at 310 K through
the Langevin thermostat and pressure was kept at 1 atm with the
Langevin Nosé-Hoover barostat[99,100] as implemented in NAMD[96].
To constrain water molecule geometries and bond vibrations, SETTLE
and RATTLE algorithms were applied, respectively[101,102]. Electrostatics at long range were treated using PME[103], using a smooth 12
Å cutoff for short-range electrostatics. From the resulting trajectory,
ten random snapshots were taken per system and refined in a QM/MM
MD simulation with the NAMD QM/MM interface[104]. The HF-3c
method[93] as implemented in Orca 4[69,94] was used for the quantum
region which consisted of Lf. Point charges of the MM region were
included via electrostatic embedding up to 12 Å distance from the

quantum region. The QM/MM systems were first minimized for 100 steps and subsequently propagated for 12 ps at a 0.5 fs time step. Temperature and pressure were maintained as explained above. No constraints were applied to the simulation systems. This setup resulted in ten individual QM/MM trajectories per system from which ten random snapshots were extracted each with MDAnalysis[105,106]. Embedding potentials for every extracted snapshot, i.e., 100 per solvent system, were generated with PyFraME[86] including solvent molecules within 15 Å proximity. It has been shown in previous work that this distance cutoff is sufficient for convergence of polarization effects in PE approaches[107]. Multipoles through second order and polarizabilities were obtained at the PBE0/cc-pVDZ level of theory as implemented in the Dalton program[63] using the LoProp approach[87,89]. For each final snapshot, the five lowest singlet excited states were calculated using PE-ADC(2)/cc-pVDZ with perturbative excitation energy corrections. An ordinary least-squares (OLS) fit of the ptSS and ptLR corrections versus change in dipole moment squared, $\Delta\mu^2$, and oscillator strength (length gauge), f, respectively, was performed using the equations

$$\Delta E_{\text{ptLR}} = -\alpha f \qquad (5.1)$$

for the ptLR correction energy and

$$\Delta E_{\text{ptSS}} = -\alpha \Delta\mu^2 \qquad (5.2)$$

for the ptSS correction energy. The fitting parameters, α, are related to the effective polarizability of the environment, neglecting pre-factors and dimensions of the fit. However, this is not important since in the analysis, only ratios of fitting parameters are taken into account. The OLS fits were calculated including all five excited states of all snapshots.

5.3 Lumiflavin Embedded in Dodecin Environment

From an MD trajectory obtained in previous work[47], ten snapshots for subsequent refinement with QM/MM MD simulations were extracted.

One Lf ligand and its adjacent tryptophan residue (W36) were defined
as the quantum region. The quantum region is depicted in Figure 5.2b.
Bonds between the quantum and classical region in the W36 backbone
were treated as explained in Ref. 47. To prevent over-polarization
at the QM/MM boundary, the charge shifting (CS) scheme[108] was
applied to shift charges adjacent to QM/MM bonds. Point charges
of the MM region were included via electrostatic embedding within a
12 Å distance cutoff. As explained before, NAMD 2.12 was used for
the MM region, whereas the QM region was treated with HF-3c[93].
Electrostatics, temperature, and pressure were handled as explained
above. The QM/MM MD simulation was run for 12 ps at a 0.5 fs
integration time step after 100 minimization steps.

From the ten QM/MM MD simulations, ten snapshots were extracted per trajectory, and embedding potentials were generated for the
extracted snapshots using PyFraME[86]. The dodecin environment
setup is graphically illustrated in Figure 5.2a. The core quantum
region in the PE-ADC calculation (Fig. 5.2b) consisted of Lf and
the adjacent W36 residue, which will be referred to as Lf-W36 in
the following. The protein was parametrized using PBE0/cc-pVDZ
for multipole moments up to second order and polarizabilities using the LoProp approach. All ions within a distance of 50 Å from
the core were taken into account through the formal charge of the
ion and isotropic polarizabilities calculated with Dalton[63] at the
B3LYP/aug-cc-pVTZ level of theory.[83,109] Water molecules within
a 10 Å radius were parametrized using LoProp including multipole
moments through quadrupoles and dipole-dipole polarizabilities. From
10 Å to 20 Å, water molecules were modeled with averaged charges and
isotropic polarizabilities[110]. TIP3P water was further used to model
the solvent between 20 Å and 50 Å. To avoid over-polarization, site
polarizabilities were removed in 1.2 Å distance of the quantum region,
and charges of these sites were redistributed to the nearest neighbor
site[111]. Higher-order multipole moments of the redistributed sites
were removed. The energetically lowest five singlet excited states were
obtained using PE-ADC(2)/cc-pVDZ while keeping the core and the
highest 70 virtual orbitals frozen, following the restricted virtual space

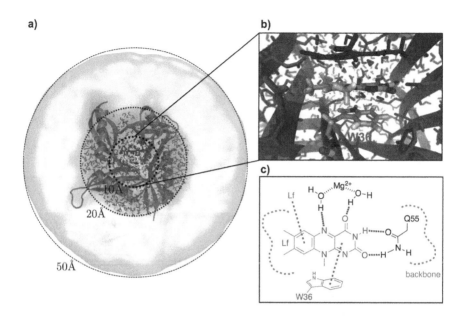

Figure 5.2: Dodecin environment setup for PE-ADC(2). **a)** Solvent shells in parametrization for dodecin PE-ADC calculation. Inner solvent shell ($r \leq 10$Å, dark blue) was parametrized with LoProp. Middle solvent shell (10Å $< r \leq 20$Å, orange) uses pre-defined charges and isotropic polarizabilities. Outer solvent shell (20Å $< r \leq 50$Å, white surface) uses TIP3P water. Protein (cyan) is fully parametrized with LoProp. Ions are included up to 50 Å distance and have pre-defined charges and isotropic polarizabilities. Ions are not shown for simplicity. **b)** Zoom-in of the dodecin active site. **c)** Interaction diagram. Purple dashed lines: hydrogen bonds, green dashed lines: π-stacking. The quantum region for PE-ADC calculation is drawn in magenta.[1]

(RVS) ADC scheme[112]. To further investigate the contribution of the environment to the possible CT, ADC(2)/cc-pVDZ calculations of the isolated Lf-W36 system, i.e., without PE, were further performed on snapshots with a ptSS correction smaller than -0.05 eV for the first excitation.

6 Results and Discussion

PE-ADC has been employed to perform a set of case studies, ranging from the small water-coordinated chromophore, PNA, over the medium-sized Lf molecule in bulk solvent to a protein system that embeds Lf. The benchmark studies probe different physical aspects of the PE-ADC method. In this regard, $n\pi^*$, $\pi\pi^*$, as well as inter- and intramolecular CT excitations are benchmarked. The computational results presented in the following have already been published in Ref. 113.

6.1 para-Nitroaniline in Presence of Water Clusters

Two different types of electronic excitations have been benchmarked in the case of PNA (Fig. 6.1) using PE-ADC(2) and PE-ADC(3), i.e., the energetically lowest singlet $n\pi^*$ and $\pi\pi^*$ excitations, in the presence of small water clusters. PNA forms hydrogen bonds with the surrounding water molecules via the nitro (NO_2) group in the subset of structures studied here. The NO_2 group acts as a hydrogen bond acceptor via the oxygen lone pairs. All results for PNA excitations using ADC(2) and ADC(3) calculations in vacuum, super-system, and PE are summarized in Tables 6.1 and 6.2. Even though hydrogen bonds between the solvent and solute exist, neither over-polarization nor electron spill-out [114] was observed. The ground and excited state dipole moments are displayed in Figure 6.1a, and the detachment-attachment densities for both states of the structure including six water molecules obtained within an ADC(2) super-system calculation are shown in Figure 6.1b. For the lowest state, which has $n\pi^*$ character,

© The Editor(s) (if applicable) and The Author(s), under exclusive license to
Springer Fachmedien Wiesbaden GmbH, part of Springer Nature 2020
M. Scheurer, *Polarizable Embedding for the Algebraic- Diagrammatic
Construction Scheme*, BestMasters, https://doi.org/10.1007/978-3-658-31281-7_6

a blue-shift in the presence of water is expected because the dipole moment of PNA is reduced upon excitation (Fig. 6.1, Tabs. 6.1 and 6.2). This should result in a destabilization of the excited state[115,116]. The blue-shift can also be rationalized by the detachment of electron density at the lone pair of the nitro group, destabilizing hydrogen bonds with coordinated water molecules (Fig. 6.1b). Since the change in density upon excitation as well as the oscillator strength of $S_0 \rightarrow S_1$ are rather small, minor contributions through perturbative corrections are to be expected.

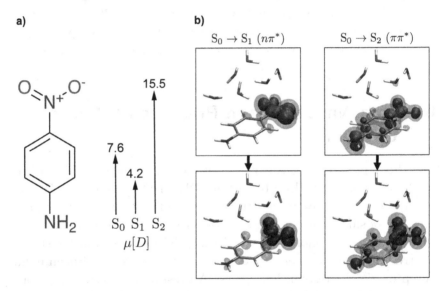

Figure 6.1: **a)** Lewis structure and dipole moments (arrows) of PNA for different electronic states in vacuum. **b)** Detachment (red) and attachment (blue) densities of the two lowest excitations of PNA in the presence of six water molecules. The $n\pi^*$ excitation is local in nature, where electron density is transferred from the lone pair to the π^* orbital of the nitro group. Hence, the proximal hydrogen bond is weakened upon photoexcitation. An intramolecular charge-transfer takes place upon $\pi\pi^*$ excitation, dramatically increasing the dipole moment of PNA.[1]

Table 6.1: PE-ADC(2) results for PNA singlet excitations in presence of n H_2O molecules.[a]

n		\multicolumn{3}{c}{$n\pi^*$}	\multicolumn{3}{c}{$\pi\pi^*$}
		vacuum	s.-s.[b]	PE	vacuum	s.-s.	PE
2	E [eV]	3.827	3.923	3.930	4.599	4.299	4.328
	f	0	0	0.000	0.453	0.436	0.447
	$\Delta\mu$ [D]	-3.38	-3.182	-3.214	7.882	8.611	8.519
	ptSS [eV]	-	-	-0.001	-	-	-0.010
	ptLR [eV]	-	-	-0.000	-	-	-0.005
4	E [eV]	3.828	3.884	3.894	4.597	4.386	4.402
	f	0	0	0.000	0.454	0.433	0.451
	$\Delta\mu$ [D]	-3.383	-3.327	-3.282	7.862	8.264	8.332
	ptSS [eV]	-	-	-0.002	-	-	-0.010
	ptLR [eV]	-	-	-0.000	-	-	-0.005
6	E [eV]	3.829	3.939	3.966	4.6	4.295	4.367
	f	0	0	0.000	0.454	0.427	0.445
	$\Delta\mu$ [D]	-3.383	-3.117	-3.155	7.862	8.638	8.435
	ptSS [eV]	-	-	-0.002	-	-	-0.018
	ptLR [eV]	-	-	-0.000	-	-	-0.011

[a] $n\pi^* = 1^1A_2$, $\pi\pi^* = 1^1A_1$ for gas-phase optimized geometries with point group symmetry.
[b] s.-s.: super-system
$\Delta\mu$ refers to the difference dipole moment between ground state and excited state.
Perturbative corrections have already been incorporated into the PE excitation energies.

In the calculations, a consistent blue-shift in presence of water molecules compared to the *in vacuo* case is observed. This blue-shift was also reproduced by PE-ADC, albeit only through contributions of the polarizable environment to the ground state wave function. As expected, both ptSS and ptLR corrections are almost equal to zero, and the solvent blue-shift is solely modeled through the PE-HF ground state. On the contrary, the $S_0 \rightarrow S_2$ transition exhibits quite different properties. This $\pi\pi^*$ excitation corresponds to an intramolecular

Table 6.2: PE-ADC(3) results for PNA singlet excitations in presence of n H_2O molecules.[a]

n		$n\pi^*$ vacuum	$n\pi^*$ s.-s.	$n\pi^*$ PE	$\pi\pi^*$ vacuum	$\pi\pi^*$ s.-s.	$\pi\pi^*$ PE
2	E [eV]	4.082	4.175	4.177	4.555	4.308	4.335
	f	0	0	0.000	0.400	0.398	0.407
	$\Delta\mu$ [D]	-2.464	-2.237	-2.298	5.944	6.541	6.449
	ptSS [eV]	-	-	-0.000	-	-	-0.007
	ptLR [eV]	-	-	-0.000	-	-	-0.005
4	E [eV]	4.083	4.146	4.148	4.551	4.379	4.392
	f	0	0	0.000	0.401	0.391	0.407
	$\Delta\mu$ [D]	-2.466	-2.43	-2.385	5.935	6.229	6.307
	ptSS [eV]	-	-	-0.000	-	-	-0.006
	ptLR [eV]	-	-	-0.000	-	-	-0.005
6	E [eV]	4.084	4.203	4.218	4.554	4.318	4.376
	f	0	0	0.000	0.401	0.390	0.404
	$\Delta\mu$ [D]	-2.466	-2.168	-2.208	5.933	6.579	6.405
	ptSS [eV]	-	-	-0.001	-	-	-0.012
	ptLR [eV]	-	-	-0.000	-	-	-0.010

[a] Details see Tab. 6.1

charge-transfer, giving rise to a significant increase in dipole moment. Electron density is detached mostly from the amino group of PNA and attached to the nitro group on the opposite site of the molecule (Fig. 6.1b). In this case, a red-shift of the excitation energy is expected, since the excited state is stabilized by a polar solvent[117,118]. As can be seen in Tables 6.1 and 6.2, this holds true for the super-system calculation and is accurately captured by PE-ADC. Even though the change in dipole moment is dramatic and the oscillator strength is much larger for the CT excitation, the perturbative corrections are small compared to the zeroth-order contribution through the environment, but correct the overestimated excitation energy in the right direction. Hence, the magnitude of the perturbative corrections can be seen as an error estimate for the fact that a) the excited state wave function and the induced moments are not self-consistently obtained (ptSS) and b) no response treatment of the PE operator is directly included (ptLR). If such corrections are small, the error being made by not including the environment response in the ADC calculation is also small. These perturbative corrections always alter overestimated excitation energies in the right direction, since the correction terms are quadratic in the induced moments and carry a negative pre-factor (eqs (3.7) and (3.8)). The ptSS and ptLR corrections for the $\pi\pi^*$ excitation are largest in the presence of six water molecules and would be even higher if the number of solvent molecules were further increased[17].

The maximum absolute error observed when comparing PE-ADC with the super-system ADC calculations are as small as 0.07 eV for ADC(2) and 0.06 eV for ADC(3) due to the localization of the electronic excitations processes to the chromophore only (Fig. 6.1b). These errors are well below the intrinsic errors of the used ADC methods[36,39,119]. All in all, the expected solvent effects are accurately captured for both types of excitations in water-coordinated PNA. Also, the zeroth-order contribution just through the PE-HF ground state has been found to have the largest effect on the solvent shift, as has been shown using other approaches as well[17,29,59,60]. Over-estimated excitation energies were narrowed by ptSS corrections which are larger for significant changes in dipole moment upon excitation, whereas ptLR corrections contribute in case of a large oscillator strength.

6.2 Lumiflavin in Bulk Solvent

The following case study aims to model bulk solvent effects on the electronic excitations of Lf through PE-ADC(2). In this context, a super-system ADC calculation is not possible due to the system size. For this reason, bulk solvent effects from water and cyHex to PCM-ADC(2) calculations will be compared. The latter is well-established and has proven a reliable technique to obtain excitation energies in bulk solvent[40,41].

Table 6.3: Comparison of ADC(2), PE-ADC(2) and PCM-ADC(2) excitations of Lf in water and cyHex solvent.[a,b]

$S_0 \to S_i$[c]		vacuum	H_2O		cyHex	
			PE	PCM	PE	PCM
S_1 ($\pi\pi^*$)	E [eV]	3.230	3.073	3.070	3.129	3.122
	f	0.257	0.243	0.262	0.238	0.257
	$\Delta\mu$ [D]	1.901	1.950	2.312	2.009	2.074
S_2 ($n\pi^*$)	E [eV]	3.610	3.579	3.773	3.565	3.663
	f	0.001	0.005	0.001	0.003	0.001
	$\Delta\mu$ [D]	-4.841	-3.490	-2.844	-3.637	-3.59
S_3 ($n\pi^*$)	E [eV]	3.804	3.842	3.971	3.842	3.787
	f	0.000	0.002	0.000	0.001	0.000
	$\Delta\mu$ [D]	-6.224	-6.272	-8.521	-6.706	-8.021
S_4 ($\pi\pi^*$)	E [eV]	4.498	4.245	4.066	4.301	4.247
	f	0.121	0.139	0.161	0.129	0.138
	$\Delta\mu$ [D]	6.581	6.315	8.341	6.574	7.546
S_5 ($n\pi^*$)	E [eV]	4.599	4.568	n.c.	4.546	n.c.
	f	0.000	0.030	n.c.	0.013	n.c.
	$\Delta\mu$ [D]	-4.700	-5.588	n.c.	-5.707	n.c.

[a] Median values of corrected excitation energies and properties are shown for PE-ADC(2).
[b] n.c. = not converged.
[c] Character of the transition in parentheses.

For PE-ADC(2), solute-solvent configurations were sampled by means of QM/MM MD simulations, whereas gas-phase ADC(2) and PCM-ADC(2) excitation energies were obtained from optimized structures. The results including excitation energies, oscillator strengths, and changes in dipole moment upon excitation are shown in Table 6.3. Characterization of the excitations was achieved using natural transition orbitals (NTOs) depicted in Figure 6.2. The solvent shift

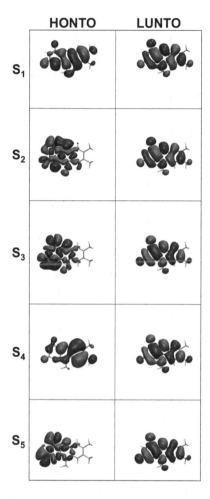

Figure 6.2: Natural transition orbitals of Lf singlet excitations in vacuum. The highest occupied NTO (HONTO) and the lowest unoccupied NTO (LUNTO) for each state are depicted.[1]

will be referred to as the difference between vacuum and solvent excitation energies. For the lowest, bright $\pi\pi^*$ excitation, PE and PCM results give almost identical results. The expected red-shift is present for both solvents, where water shows a slightly larger red-shift than cyHex for both embedding schemes. Predicted solvent shifts for the S_2 $n\pi^*$ dark state differ between PCM-ADC(2) and PE-ADC(2): PE predicts a red-shift of the excitation energy, whereas PCM shows a blue-shift. However, the red-shift is small and could simply result from the sampled geometries underlying the PE-ADC(2) calculations, whereas an optimized structure is used in the PCM case. This hypothesis would, nevertheless, have to be further investigated. Differing solvent shifts are also observed for the third singlet state, which is an $n\pi^*$ excitation as well. A $\pi\pi^*$ excitation with CT character corresponds to the fourth excited state of Lf. In this case, a red-shift is again correctly predicted by PE and PCM. This red-shift is more pronounced in water solvent, where PCM predicts a shift approximately 0.2 eV larger than PE. The comparison of PE and PCM excitation energies and excited state properties shows a consistent behavior for spectroscopically relevant bright states, i.e., the $S_0 \to S_1$ and $S_0 \to S_4$ transitions in this case. The trends for solvent shifts through bulk solvation modeled by PE-ADC(2) are thus comparable to PCM-ADC(2) which is a thoroughly benchmarked method for these applications[40].

To gain further insight into the excitation energies and perturbative corrections, a statistical analysis of the ADC calculations is needed. Accordingly, the PE-ADC(2) results for corrected excitation energies, ptSS, and ptLR corrections are depicted as box plots in Figure 6.3. The box plots show the ranges of corrected excitation energies resulting from the configurational sampling. A minor amount of outliers is observed here for all states except for S_2. However, the energy range spanned by the boxes (50% of the observations) is rather narrow. Higher-lying states cover a broader energy range for ptSS correction terms, whereas ptLR corrections are very localized. Thus, the ptSS corrections term seems to be rather sensitive to the underlying geometry. Generally, the box plots show a rather condensed representation of the manifold of PE-ADC excited state calculations, and are thus the ideal tool to assess the geometrical sampling for the final excited state calculations. In case of highly fluctuating excitation energies, one would have to improve the sampling, as explained in the general work

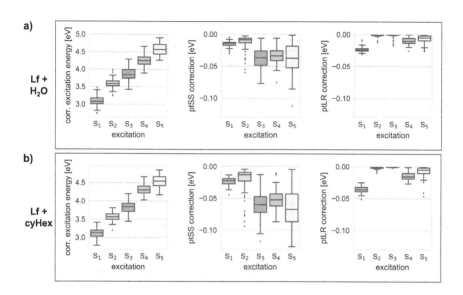

Figure 6.3: PE-ADC(2) excitation energies and corrections for Lf in **a)** water and **b)** cyHex solvent. Corrected excitation energies and ptSS and ptLR contributions are depicted as box plots for the five lowest excited singlet states of Lf. The boxes comprise 50% of the observed data, where the antennae mark lower and upper boundaries of 1.5 times inter-quartile range (IQR). Data points outside 1.5 IQR are defined as outliers (diamonds).[1]

flow (Fig. 1). As discussed before, the total solvent red-shift on $\pi\pi^*$ excitations is larger for water than for cyHex. On the contrary, both the ptSS and ptLR corrections display a larger magnitude in cyHex solvent compared to water. The interpretation of the perturbative corrections in this context will be elaborated by correlating the ptSS and ptLR terms with the squared change in dipole moment and the oscillator strength, respectively, because these properties are inherently linearly related to the correction terms through equations (3.7) and (3.8). Corresponding linear fits, yielding what is here called the 'effective' solvent polarizability as a fitted parameter, are plotted in

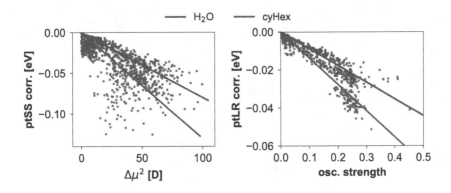

Figure 6.4: Correlation between perturbative corrections and excited
state/transition properties. For both water and cyHex
solvent, ptSS and ptLR corrections are shown vs. squared
difference dipole moment ($\Delta\mu^2$) and oscillator strength, respectively, for all states.[1]

Figure 6.4. Detailed fitting results are shown in Table 6.4. The ratios

Table 6.4: OLS fit parameters for α

solvent	ΔE	α	R^2
water	ptSS	0.0008 eV/D^2	0.761
water	ptLR	0.0881 eV	0.966
cyHex	ptSS	0.0013 eV/D^2	0.731
cyHex	ptLR	0.1382 eV	0.973

between the effective polarizabilities of cyHex and water solvent are
1.63 for ptSS and and 1.57 for ptLR. That is, the cyHex environment
is effectively more polarizable than the water environment, leading to
larger perturbative energy correction terms. Using the configurational
sampling through QM/MM MD, the 'effective' solvent polarizabilities
are easily obtained directly from fitting the perturbative energy corrections. All in all, this analysis nicely illustrates the general relationship
between excited state properties and the expected magnitude of ptSS
and ptLR corrections.

6.3 Protein Environment Effects on the Charge-Transfer Excitation in Dodecin

To showcase the capabilities of PE-ADC for addressing excited states of chromophores embedded in large, complex environments, I here investigate the CT excitation involved in the photoprotection mechanism of the flavin-binding protein dodecin[120–122]. Dodecin stores flavin derivatives and efficiently photoprotects them by a multi-step quenching mechanism[47]. The initial key step in the mechanism most likely comprises a CT excitation, where an electron is transferred from a tryptophan residue (W36) to Lf, forming a positively charged tryptophan radical and a negatively charged flavosemiquinone radical. In previous work[47], the CT excitation of Lf-W36 was modeled using ADC(2) calculations *in vacuo*. Since such a CT excitation creates two oppositely charged fragments, it exhibits a large change in electron density. Not only is the system challenging due to the nature of the excited state, but also with respect to the heterogeneous interactions of the core region with the environment: Lf forms hydrogen bonds with solvent water molecules and a bifurcated hydrogen bond with Q55 in the protein. Further, it interacts with another Lf molecule by π-stacking. The heterogeneity of the core-environment interactions modeled through PE is depicted in Figure 5.2c. Obviously, these interactions are impossible to capture using a CSM. PE-ADC(2) is also found to be computationally affordable on the Lf-W36 system since the calculations only require additional time during the SCF procedure compared to ADC(2) calculations without PE (data not illustrated).

The results for PE-ADC(2) excitation energies and perturbative corrections for the three energetically lowest singlet states are presented in Figure 6.5. The two lowest states are almost degenerate, as observed in the *in vacuo* case[47], while the third excited state covers a spectrum of almost 2 eV among all snapshots. The observed similarity between the lowest two states can further be examined with the dominant amplitudes involved in the transition (Fig. 6.6). In fact, S_1 and

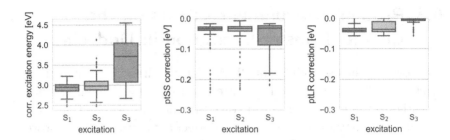

Figure 6.5: PE-ADC(2) excitation energies and corrections for Lf-W36
in dodecin environment. The presented box plots were prepared as for Fig. 6.3.[1]

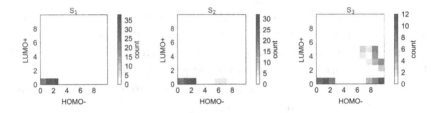

Figure 6.6: Heat map of MO contributions with largest amplitudes for
PE-ADC(2) on Lf-W36. For the three energetically lowest
excited states, the occurrences of MO transition amplitudes
are shown from all sampled snapshots. The two lowest states
share the same main MO contributions, whereas the third
lowest state shows a broader pattern in dominating transition amplitudes, including lower-lying occupied and higher-
lying virtual orbitals.[1]

S_2 show an almost identical pattern, including HOMO-2→LUMO,
HOMO-1→LUMO, and HOMO→LUMO as dominant transitions.
However, S_3 only contains these amplitudes to some extent in addition
to other major orbital contributions. The HOMO and LUMO of a
representative snapshot are depicted in Figure 6.7. Clearly, the HOMO
and LUMO are localized on the W36 and Lf, respectively, giving rise to
a CT from W36 to Lf. The other dominant amplitudes show a mixture

of CT and local excitation (LE)[47] (data not illustrated), where HOMO→LUMO exhibits the most prominent fragment separation. Since the energy range of the third singlet excitation also covers S_1 and

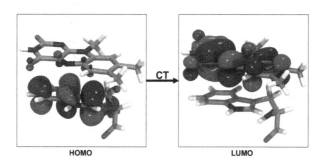

Figure 6.7: HOMO to LUMO transition comprising a possible Lf-W36 charge-transfer.[1]

S_2 excitation energies, it is possible that in a few snapshots, $S_0 \rightarrow S_3$ corresponds to a CT excitation. This is also corroborated by the aforementioned orbital contributions.

The ptSS correction energies are localized around -0.03 eV and -0.2 eV for the first two states, and a broader range of values is covered by the data for the third state. The data for ptLR correction energies contain very few outliers for S_1 and S_2 with a median of -0.039 eV and -0.036 eV, respectively. The median ptLR correction for S_3 is -0.003 eV, thus, almost equal to zero. Accordingly, ptSS corrections for S_3 and the outliers clustered around -0.2 eV for S_1 and S_2 are dominating over ptLR corrections in the Lf-W36 system.

For a CT state, the ptSS correction is expected to be large due to the charge separation between the two fragments and the resulting change in dipole moment upon excitation. To further unravel the effect of the polarizable environment on the CT character, snapshots that showed a ptSS correction smaller than -0.05 eV for the first excited state were re-calculated without PE, i.e., on the isolated Lf-W36 system only. These snapshots should supposedly exhibit a strong CT character due to the large magnitude of the ptSS correction. To

avoid visual inspection of all snapshots, statistical analysis of the CT
character was performed using the electron-hole distance $d_{h\to e}$ [35,90,91]
which is obtained from the excitonic wave function. It quantifies a
CT through the distance of the hole and electron centroid and is
thus ideal to characterize a large amount of ADC calculations. A
CT results in a large shift in electron density from one fragment to
another. This should result in a large increase in dipole moment upon
excitation and consequently a large ptSS energy correction. Figure 6.8a
shows the expected relationship between ptSS correction and electron-
hole distance: The larger the electron-hole distance, the larger the
magnitude of the ptSS correction. Interestingly, both quantities are
related to different properties of the electronic excitations: electron-
hole distances are obtained from the excitonic wave function, i.e., from
the transition density matrix, whereas the ptSS term originates from
the difference density matrix. Nevertheless, the electron-hole distance
quantifies a charge transfer through electronic excitation which is also
reflected by the ptSS correction term. To quantify the environmental
effect on the CT character, electron-hole distances for the extracted
snapshots with and without PE are compared in Figure 6.8b. With
PE, S_1 electron-hole distances are clustered above 1 Å, whereas the
values for S_2 and S_3 are almost evenly distributed between 1 Å and 3
Å. The corresponding median values are 2.81 Å, 2.33 Å, and 1.17 Å,
respectively. This picture is dramatically changed if no environment
through PE is included for the same snapshots: For all states, electron-
hole distances are clustered around 1 Å with the median values 1.07
Å, 0.97 Å, and 0.56 Å, respectively. As a consequence, the electron-
hole separation is diminished without the polarizable environment,
increasing the LE character of the excited states. Clearly, not all
observed electronic excitations are pure CT states, but rather a mixture
of CT and LE. The presence of the environment, however, increases the
CT character of the low-lying excited states. In that way, the protein
environment promotes CT excitations to trigger the photoprotection
mechanism. A local excitation on the Lf fragment would not result in
subsequent proton transfer and follow-up excited state quenching [47].
From a biochemical point of view, the protein environment seems to be

Obviously, f is proper, convex, lower semicontinuous and $\mathbb{R}_+^n$-increasing on $F^1(\mathrm{dom}\, F^1) + K_0 \subseteq \mathbb{R}_+^n$, the function F^1 is proper, $\mathbb{R}_+^n$-epi closed and $\mathbb{R}_+^n$-convex as well as $0_{\mathbb{R}^n} \in \mathrm{ri}(F^1(\mathrm{dom}\, F^1) - \mathrm{dom}\, f + K_0) = \mathbb{R}^n$ and thus, it follows by Theorem 3.6 that

$$\gamma_C^*(x_1^*, ..., x_n^*) = \min_{\substack{y_i^{0*} \in \mathbb{R}_+, \\ i=1,...,n}} \{f^*(y_1^{0*}, ..., y_n^{0*}) + ((y_1^{0*}, ..., y_n^{0*})^T F^1)^*(x_1^*, ..., x_n^*)\}.$$

From (4.3) we have for $a_i = 0$, $i = 1, ..., n$, that

$$
f^*(y_1^{0*}, ..., y_n^{0*}) = \begin{cases} 0, & \text{if } \sum_{i=1}^n \lambda_i \leq 1, \ \lambda_i \geq 0, \ y_i^{0*} \leq \lambda_i, \ i = 1, ..., n, \\ +\infty, & \text{otherwise}, \end{cases}
$$

$$
= \begin{cases} 0, & \text{if } \sum_{i=1}^n y_i^{0*} \leq 1, \ y_i^{0*} \geq 0, \ i = 1, ..., n, \\ +\infty, & \text{otherwise}. \end{cases} \tag{4.22}
$$

In addition, it holds

$$((y_1^{0*}, ..., y_n^{0*})^T F^1)^*(x_1^*, ..., x_n^*) = \sum_{i=1}^n \sup_{x_i \in Y_i} \{\langle x_i^*, x_i \rangle - y_i^{0*} \gamma_{C_i}(x_i)\} = \sum_{i=1}^n (y_i^{0*} \gamma_{C_i})^*(x_i^*).$$

For $y_i^{0*} > 0$ holds

$$
(y_i^{0*} \gamma_{C_i})^*(x_i^*) = \begin{cases} 0, & \text{if } \gamma_{C_i^0}(x_i^*) \leq y_i^{0*}, \\ +\infty, & \text{otherwise}, \end{cases}
$$

and if $y_i^{0*} = 0$, then

$$
(0 \cdot \gamma_{C_i})^*(x_i^*) = \sup_{x_i \in Y_i} \{\langle x_i^*, x_i \rangle\} = \begin{cases} 0, & \text{if } x_i^* = 0_{Y_i^*}, \\ +\infty, & \text{otherwise}. \end{cases}
$$

This implies that

$$((y_1^{0*}, ..., y_n^{0*})^T F^1)^*(x_1^*, ..., x_n^*) = \sum_{i=1}^n (y_i^{0*} \gamma_{C_i})^*(x_i^*)$$

$$
= \begin{cases} 0, & \text{if } \gamma_{C_i^0}(x_i^*) \leq y_i^{0*}, \ i = 1, ..., n, \\ +\infty, & \text{otherwise}, \end{cases} \tag{4.23}
$$

and hence, one has by (4.22) and (4.23)

$$
\gamma_C^*(x^*) = \begin{cases} 0, & \text{if } \sum_{i=1}^n y_i^{0*} \leq 1, \ y_i^{0*} \geq 0, \ \gamma_{C_i^0}(x_i^*) \leq y_i^{0*}, \ i = 1, ..., n, \\ +\infty, & \text{otherwise}. \end{cases}
$$

$$
= \begin{cases} 0, & \text{if } \sum_{i=1}^n \gamma_{C_i^0}(x_i^*) \leq 1, \\ +\infty, & \text{otherwise}. \end{cases}
$$

For $\lambda > 0$ it follows

$$(\lambda\gamma_C)^*(x^*) = \lambda\gamma_C^*\left(\frac{1}{\lambda}x^*\right) = \begin{cases} 0, & \text{if } \sum_{i=1}^n \gamma_{C_i^0}(x_i^*) \leq \lambda, \\ +\infty, & \text{otherwise.} \end{cases} \tag{4.24}$$

Moreover, by (4.20) follows for $\lambda = 0$

$$(0 \cdot \gamma_C)^*(x^*) = \sum_{i=1}^n \sup_{x_i \in Y_i} \{\langle x_i^*, x_i\rangle\} = \begin{cases} 0, & \text{if } x_i^* = 0_{Y_i^*} \ \forall i = 1, ..., n, \\ +\infty, & \text{otherwise.} \end{cases} \tag{4.25}$$

and as $\sum_{i=1}^n \gamma_{C_i^0}(0_{Y_i^*}) = \sum_{i=1}^n \sup\{\langle 0_{Y_i^*}, x_i\rangle\} = 0$, we have by (4.24) and (4.25) for the Lagrange dual problem

$$(\widetilde{D}_L^{\gamma^0}) \quad \sup_{\lambda \geq 0}\left\{-\lambda - (\lambda\gamma_C)^*(x^*)\right\} = \sup_{\lambda \geq 0}\left\{-\lambda : \sum_{i=1}^n \gamma_{C_i^0}(x_i^*) \leq \lambda\right\}.$$

It is obvious that the Slater constraint qualification for the primal-dual problem $(\widetilde{P}^{\gamma^0}) - (\widetilde{D}_L^{\gamma^0})$ is fulfilled and thus, strong duality holds, i.e.,

$$\gamma_{C^0}(x^*) = \sup_{x \in C}\{\langle x^*, x\rangle\} = \min_{\lambda \geq 0}\left\{\lambda : \sum_{i=1}^n \gamma_{C_i^0}(x_i^*) \leq \lambda\right\} = \sum_{i=1}^n \gamma_{C_i^0}(x_i^*).$$

$\square$

4.1.2 Perturbed minimal time functions

Given a nonempty set $\Omega \subset X$ and a proper function $f : X \to \overline{\mathbb{R}}$, we define the *extended perturbed minimal time function* $\mathcal{T}_{\Omega,f}^C : X \to \overline{\mathbb{R}}$ as the infimal convolution of γ_C, f and δ_Ω, i.e. $\mathcal{T}_{\Omega,f}^C := \gamma_C \square f \square \delta_\Omega$, more precisely

$$\mathcal{T}_{\Omega,f}^C(x) := \inf_{y \in X, \ z \in \Omega}\{\gamma_C(x - y - z) + f(y)\}.$$

Remark 4.6. *To the best of our knowledge the function $\mathcal{T}_{\Omega,f}^C$ has not been considered in this form in the literature yet and it covers as special cases several important functions. For instance, if $f = \delta_{\{0_X\}}$, then one gets the classical minimal time function (see [78–80]) $\mathcal{T}_\Omega^C : \mathbb{R}^n \to \overline{\mathbb{R}}$, $\mathcal{T}_\Omega^C(x) := \inf\{t \geq 0 : (x - tC) \cap \Omega \neq \emptyset\}$, that, when $\Omega = \{0_X\}$ collapses to the gauge function. In [76] one finds two perturbations of the classical minimal time function, $\gamma_C \square f$ (introduced in [108] and motivated by a construction specific to differential inclusions) and $\gamma_C \square (f + \delta_\Omega)$, that contains as a special case the perturbed distance function introduced in [98]. The latter function has motivated us to introduce $\mathcal{T}_{\Omega,f}^C$, where the function f and the set Ω do not share the same variable anymore and can thus be split in the dual representations. Other generalizations of the classical minimal time function can*

be found, for instance, in [28, 84, 85]. The minimal time function and its generalizations
have been employed in various areas of research such as location theory (cf. [78, 85]),
nonsmooth analysis (cf. [28, 76, 79, 80, 84, 85, 98, 108]), control theory and Hamilton-Jacobi
partial differential equation (mentioned in [28]), best approximation problems (cf. [98])
and differential inclusions (cf. [108]). Note also the connection observed in [85] between
the minimal time function and the scalarization function due to Tammer (Gerstewitz)
considered in vector optimization.

Moreover, as $\mathcal{T}_{\Omega,f}^C$ is an infimal convolution its conjugate function turns into (see [14,
Proposition 2.3.8.(b)]) $(\mathcal{T}_{\Omega,f}^C)^* = \gamma_C^* + f^* + \sigma_\Omega = \delta_{C^0} + f^* + \sigma_\Omega$ (see Remark 4.5).

Since $\operatorname{dom}\gamma_C \neq \emptyset$ and γ_C is a nonnegative function, it follows by [14, Lemma 2.3.1.(b)]
that $\gamma_C^* = \delta_{C^0}$ is proper, convex and lower semicontinuous. If f has an affine minorant,
then f^* is a proper, convex and lower semicontinuous function, too. Further, under the
additional assumption $C^0 \cap \operatorname{dom} f^* \cap \operatorname{dom}\sigma_\Omega \neq \emptyset$, [14, Theorem 2.3.10] yields that the
biconjugate of $\mathcal{T}_{\Omega,f}^C$ is, under the mentioned hypotheses, given by $(\mathcal{T}_{\Omega,f}^C)^{**} = \overline{\gamma_C^{**}\Box f^{**}\Box\sigma_\Omega^*}$,
and one can derive as byproducts conjugate and biconjugate formulae for the classical
minimal time function and its other extensions mentioned in Remark 4.6.

Theorem 4.2. *Let C be convex, closed and contain 0_X, Ω be closed and convex and
$f : X \to \overline{\mathbb{R}}$ be also convex and lower semicontinuous such that $C^0 \cap \operatorname{dom} f^* \cap \operatorname{dom}\sigma_\Omega \neq \emptyset$.
Suppose that one of the following holds*

(a) $\operatorname{epi}\gamma_C + \operatorname{epi} f + (\Omega \times \mathbb{R}_+)$ is closed,

*(b) there exists an element $x^* \in C^0 \cap \operatorname{dom} f^* \cap \operatorname{dom}\sigma_\Omega$ such that two of the functions δ_{C^0}, f^*
and σ_Ω are continuous at x^*.*

Then $\mathcal{T}_{\Omega,f}^C$ is proper, convex and lower semicontinuous and moreover, it holds

$$\mathcal{T}_{\Omega,f}^C(x) = \min_{y\in X,\, z\in\Omega} \{\gamma_C(x - y - z) + f(y)\} \ \forall x \in X,$$

i.e. the infimal convolution of γ_C, f and δ_Ω is exact .

Proof. As C is closed and convex such that $0_X \in C$, it follows by [102, Theorem 1] that
γ_C is proper, convex and lower semicontinuous. Further, the nonemptiness, closedness and
convexity of Ω imply the properness, convexity and lower semicontinuity of δ_Ω. Hence,
one gets from the Fenchel-Moreau Theorem that $\gamma_C^{**} = \gamma_C$, $f^{**} = f$ and $\delta_\Omega^{**} = \delta_\Omega$ and
from [14, Theorem 3.5.8.(a)], taking into consideration that the conjugate functions of f, δ_Ω
and γ_C are proper, convex and lower semicontinuous (as noted above), follows the desired
statement. $\qquad\Box$

Remark 4.7. *Under the hypotheses of Theorem 4.2 the subdifferential of $\mathcal{T}_{\Omega,f}^C$ can be written
for any $x \in X$ as $\partial\mathcal{T}_{\Omega,f}^C(x) = \partial\gamma_C(x - y - z) \cap \partial f(y) \cap N_\Omega(z)$, where y and z are the points
where the minimum in the definition of the infimal convolution is attained. Note, moreover,
that the subdifferential of γ_C at any $x \in X$ coincides with the face of C^0 exposed by x
(cf. [72]), i.e. $\partial\gamma_C(x) = \{x^* \in C^0 : \langle x^*, x\rangle = \sigma_{C^0}(x)\}$.*

Remark 4.8. *Results connected to the situation when the regularity condition (b) in Theorem 4.2 is fulfilled can be found, for instance, in [90]. The support function of a compact convex set is real valued and continuous, however, in the absence of compactness it is surely continuous only on the (relative) interior of its domain, while an indicator function is continuous over the interior of the corresponding set.*

Taking f to be a gauge or an indicator function one obtains the following geometrical interpretations of the generalization of the extended perturbed minimal time function.

Remark 4.9. *Let $C, \Omega, G \subseteq X$ be convex and closed sets such that $0_X \in C \cap G$. Then*

$$\mathcal{T}_{\Omega,\gamma_G}^{-C}(x) = \inf_{\substack{\alpha,\beta>0,\ z\in\Omega,\ y\in X, \\ x-y-z\in-\alpha C,\ y\in\beta G}} \{\alpha + \beta\} = \inf_{\substack{\alpha,\beta>0,\ z\in\Omega,\ k\in X, \\ x-k\in-\alpha C,\ k-z\in\beta G}} \{\alpha + \beta\} = \inf_{\substack{\alpha,\beta>0, \\ (x+\alpha C)\cap(\Omega+\beta G)\neq\emptyset}} \{\alpha + \beta\}.$$

(4.26)

The last formula suggests interpreting α as the minimal time needed for the given point x to reach the set Ω along the constant dynamics $-C$, while Ω is moving in direction of x with respect to the constant dynamics characterized by the set G. The value β gives then the minimal time needed for Ω to reach x.

Remark 4.10. *Let $S \subseteq X$, Ω and C be convex and closed, with $S \neq \emptyset$ and $0_X \in C$. Then*

$$\mathcal{T}_{\Omega,\delta_S}^{-C}(x) = \inf\{\lambda > 0 : y \in S,\ z \in \Omega,\ x - y - z \in -\lambda C\}$$
$$= \inf\{\lambda > 0 : (x + \lambda C) \cap (S + \Omega) \neq \emptyset\}.$$

(4.27)

The extended perturbed minimal time function reduces to the classical minimal time function with the target set $S + \Omega$, i.e. if the set C describes constant dynamics, then $\mathcal{T}_{\Omega,\delta_S}^{-C}(x)$ is the minimal time $\lambda > 0$ needed for the point x to reach the target set $S + \Omega$ (see for instance [78]). However, one can also write

$$\mathcal{T}_{\Omega,\delta_S}^{-C}(x) = \inf\{\lambda > 0 : y \in S,\ z \in \Omega,\ x - y - z \in -\lambda C\}$$
$$= \inf\{\lambda > 0 : (x + S + \lambda C) \cap \Omega \neq \emptyset\},$$

(4.28)

and, when C characterizes again constant dynamics, $\mathcal{T}_{\Omega,\delta_S}^{-C}(x)$ can be understood as the minimal time $\lambda > 0$ needed for the set S translated by the point x to reach the target set Ω.

Remark 4.11. When C is convex and closed with $0 \in C$, Ω is convex and compact and $f = \delta_{\{0_X\}}$, then σ_Ω is continuous and $\mathcal{T}_{\Omega,\delta_{\{0_X\}}}^C$ is proper, convex and lower semicontinuous by Theorem 4.2, and can be written as $\mathcal{T}_{\Omega,\delta_{\{0_X\}}}^C = \min_{z\in\Omega}\gamma_C(\cdot - z)$. This statement can also be found in the special case $X = \mathbb{R}^n$ in [78, Theorem 3.33 and Theorem 4.7]. Moreover, in [76] it is assumed that $\gamma_C \square (f + \delta_\Omega)$ is exact, under similar hypotheses that would actually guarantee this outcome.

Remark 4.12. *If $0_X \in \operatorname{core} C$, γ_C has a full domain, consequently so does the corresponding extended perturbed minimal time function since in general $\operatorname{dom}\mathcal{T}_{\Omega,f}^C = \operatorname{dom}\gamma_C + \operatorname{dom}f + \operatorname{dom}\delta_\Omega$.*

4.2 Single minmax location problems with perturbed minimal time functions

In this section we investigate minmax location problems where the distances are measured by perturbed minimal time functions.

Let X be a Banach space (note that most of the following investigations can be extended to a Fréchet space, too) and $a_i \in \mathbb{R}_+$, $i = 1, \ldots, n$, be given nonnegative set-up costs, where $n \geq 2$ and consider the following generalized location problem

$$(P_{h,\mathcal{T}}^S) \quad \inf_{x \in S} \max_{1 \leq i \leq n} \left\{ h_i \left(\mathcal{T}_{\Omega_i, f_i}^{C_i}(x) \right) + a_i \right\},$$

where $S \subseteq X$ is nonempty, closed and convex, $C_i \subseteq X$ is closed and convex with $0_X \in \operatorname{int} C_i$, $\Omega_i \subseteq X$ is nonempty, convex and compact, $f_i : X \to \overline{\mathbb{R}}$ is proper, convex and lower semicontinuous, $h_i : \mathbb{R} \to \overline{\mathbb{R}}$ with $h_i(x) \in \mathbb{R}_+$, if $x \in \mathbb{R}_+$, and $h_i(x) = +\infty$, otherwise, is proper, convex, lower semicontinuous and increasing on $\mathbb{R}_+$, $i = 1, \ldots, n$.

Note that the assumptions made above yield that $0_{X^*} \in C_i^0 \cap \operatorname{dom} \sigma_\Omega \cap \operatorname{dom} f_i^*$ and as $\gamma_{C_i}^* = \delta_{C^0}$ and σ_{Ω_i} are continuous functions (as $0 \in \operatorname{int} C_i$ and Ω_i is convex and compact), one gets by Theorem 4.2 that $\mathcal{T}_{\Omega_i, f_i}^{C_i}$ is a proper, convex and lower semicontinuous function with full domain and thus continuous, $i = 1, \ldots, n$. Moreover, since h_i is a proper, convex, lower semicontinuous and increasing function, $i = 1, \ldots, n$, it follows that the objective function of $(P_{h,\mathcal{T}}^S)$ is proper, convex and lower semicontinuous, which means that $(P_{h,\mathcal{T}}^S)$ is a convex optimization problem.

Now, we analyze how can be understood the location problem $(P_{h,\mathcal{T}}^S)$ in the more simple situation where the function h_i is linear continuous on $(0, +\infty)$ and f_i is the indicator function of a nonempty, closed and convex subset of X, $i = 1, \ldots, n$.

Remark 4.13. *In the context of the Remark 4.9 let us consider the following concrete minmax location problem where $h_i(x) = x + \delta_{\mathbb{R}_+}(x)$, $x \in \mathbb{R}$, $a_i = 0$, and C_i, G_i and Ω_i are closed and convex sets such that $0_X \in C_i \cap G_i$ for all $i = 1, \ldots, n$,*

$$(P_{\gamma_G,\mathcal{T}}^S) \quad \inf_{x \in X} \max_{1 \leq i \leq n} \left\{ \mathcal{T}_{\Omega_i, \gamma_{G_i}}^{-C_i}(x) \right\} = \inf_{\substack{x \in X, \ t \in \mathbb{R}, \ \mathcal{T}_{\Omega_i, \gamma_{G_i}}^{-C_i}(x) \leq t, \\ i = 1, \ldots, n}} t$$

$$= \inf_{\substack{x \in X, \ t \in \mathbb{R}, \ \inf\{\alpha_i + \beta_i > 0 : (x + \alpha_i C_i) \cap (z_i + \beta_i G_i)\} \leq t, \\ z_i \in \Omega_i, i = 1, \ldots, n}} t = \inf_{\substack{x \in X, \ \alpha_i, \ \beta_i, \ t > 0, \ \alpha_i + \beta_i \leq t, z_i \in \Omega_i, \\ (x + \alpha_i C_i) \cap (z_i + \beta_i G_i) \neq \emptyset, \ i = 1, \ldots, n}} t.$$

The last formulation allows the following economical interpretation. Given n countries, each with a growing demand G_i for a product and an average income (or average budget) characterized by the set Ω_i, $i = 1, \ldots, n$, consider a company, which produces and sells this product, planning to build a production facility. The production speed of the production facility as well as the preference of the company for a country are characterized by the sets C_i, $i = 1, \ldots, n$. Then the objective of the company is to determine a location $\bar{x}$ for a production facility such that the total demand for the product can be satisfied in the shortest

*time, i.e. the company wants to enter all lucrative markets as fast as possible. Additionally,
in the general case when the functions $h_1, \ldots, h_n$ are not necessarily linear (over their
domains), these can be seen as cost or production functions, while the set-up costs $a_1, \ldots, a_n$
are the costs of the testing of the products to meet the various specifications asked by each
of the n countries.*

Remark 4.14. *When $L_i \subseteq X$, $i = 1, \ldots, n$, are nonempty, closed and convex sets, $f_i = \delta_{L_i}$
and $h_i = \cdot + \delta_{\mathbb{R}_+}(\cdot)$, $i = 1, \ldots, n$, $(P_{h,\mathcal{T}}^S)$ reads as (see also Remark 4.10)*

$$(P_{\mathcal{T}}^S) \quad \inf_{x \in S} \max_{1 \le i \le n} \left\{ \mathcal{T}_{\Omega_i, \delta_{L_i}}^{C_i}(x) + a_i \right\} = \inf_{\substack{x \in S, \, t \in \mathbb{R}, \\ \mathcal{T}_{\Omega_i, \delta_{L_i}}^{C_i}(x) + a_i \le t, \\ i=1,\ldots,n}} t = \inf_{\substack{x \in S, \, t \in \mathbb{R}, \\ \inf\{\lambda_i > 0 : (x - \lambda_i C_i) \cap (\Omega_i + L_i) \neq \emptyset\} + a_i \le t, \\ i=1,\ldots,n}} t$$

and can be seen as finding a point $x \in S$ and the smallest number $t > 0$ such that

$$(x - (t - a_i)C_i) \cap (\Omega_i + L_i) \neq \emptyset \ \forall i = 1, \ldots, n, \tag{4.29}$$

*where C_i can be defined as a generalized ball with radius $t - a_i$, $i = 1, \ldots, n$ (see [78]).
This approach is especially useful if the target set is hard to handle, but can be split into a
Minkowski sum of two simpler sets Ω_i and L_i, $i = 1, \ldots, n$, as happens for instance with the
rounded rectangles that can be written as sums of rectangles and circles. Note also that [4]
addresses the situation when the projection onto a Minkowski sum of closed convex sets
coincides with the sums of projections into these sets. Alternatively, $(P_{\mathcal{T}}^S)$ can be written as*

$$(P_{\mathcal{T}}^S) \quad \inf_{x \in S} \max_{1 \le i \le n} \left\{ \mathcal{T}_{\Omega_i, \delta_{L_i}}^{C_i}(x) + a_i \right\} = \inf_{\substack{x \in S, \, t \in \mathbb{R}, \\ \inf\{\lambda_i > 0 : (x - L_i - \lambda_i C_i) \cap \Omega_i \neq \emptyset\} + a_i \le t, \\ i=1,\ldots,n}} t,$$

*which allows the interpretation as finding a point $x \in S$ and the smallest number $t > 0$ such
that*

$$(x - L_i - (t - a_i)C_i) \cap \Omega_i \neq \emptyset \ \forall i = 1, \ldots, n. \tag{4.30}$$

*Both (4.29) and (4.30) are generalizations of the classical Sylvester problem that consists
in finding the smallest circle that encloses finitely many given points.*

In order to approach the problem $(P_{h,\mathcal{T}}^S)$ by means of the conjugate duality concept
introduced in Chapter 3 (with $X_0 = \mathbb{R}^n$ partially ordered by the convex cone $K_0 = \mathbb{R}_+^n$,
$X_1 = X^n$ partially ordered by the trivial cone $K_1 = \{0_{X^n}\}$ and $X_2 = X$), we consider the
following functions

$$f : \mathbb{R}^n \to \overline{\mathbb{R}}, \quad f(z) := \begin{cases} \max_{1 \le i \le n} \{h_i(z_i) + a_i\}, & \text{if } z = (z_1, \ldots, z_n)^\top \in \mathbb{R}_+^n, \ i = 1, \ldots, n, \\ +\infty, & \text{otherwise,} \end{cases}$$

$$\tag{4.31}$$

$F^1 : X^n \to \mathbb{R}^n$, $F^1(y_1, \ldots, y_n) := \left(\mathcal{T}_{\Omega_1, f_1}^{C_1}(y_1), \ldots, \mathcal{T}_{\Omega_n, f_n}^{C_n}(y_n)\right)^\top$ and $F^2 : X \to X^n$, $F^2(x) :=$ $(x, \ldots, x)$. With these newly introduced functions we can write the optimization problem $(P_{h,\mathcal{T}}^S)$ as a multi-composed optimization problem

$$(P_{h,\mathcal{T}}^S) \qquad \inf_{x \in S} (f \circ F^1 \circ F^2)(x).$$

Notice that the function f is proper, convex, $\mathbb{R}_+^n$-increasing on $F^1(\mathrm{dom}\, F^1) + K_0 = \mathrm{dom}\, f = \mathbb{R}_+^n$ and lower semicontinous. Moreover, as the functions $\mathcal{T}_{\Omega_i, f_i}^{C_i}$, $i = 1, \ldots, n$, are proper, convex and lower semicontinuous, it is obvious that the function F^1 is proper, $\mathbb{R}_+^n$-convex and $\mathbb{R}_+^n$-epi-closed. In addition, as the function F^2 is linear continuous, it follows that the function F^1 does not need to be monotone (see Remark 3.5). Employing the duality concept introduced in the previous chapter we attach to $(P_{h,\mathcal{T}}^S)$ the following conjugate dual problem

$$(D_{h,\mathcal{T}}^S) \quad \sup_{\substack{z_i^{0*} \in \mathbb{R}_+, \ z_i^{1*} \in X^*, \\ i=1,\ldots,n}} \left\{ \inf_{x \in S} \left\{ \sum_{i=1}^n \langle z_i^{1*}, x \rangle \right\} - f^*(z^{0*}) - (z^{0*} F)^*(z^{1*}) \right\}, \qquad (4.32)$$

where $z^{0*} = (z_1^{0*}, \ldots, z_n^{0*})^\top \in \mathbb{R}_+^n$ and $z^{1*} = (z_1^{1*}, \ldots, z_n^{1*}) \in (X^*)^n$ are the dual variables. By Lemma 4.1 one has

$$f^*(z_1^{0*}, \ldots, z_n^{0*}) = \min_{\substack{\sum_{i=1}^n \lambda_i \leq 1, \ \lambda_i \geq 0, \\ i=1,\ldots,n}} \left\{ \sum_{i=1}^n [(\lambda_i h_i)^*(z_i^{0*}) - \lambda_i a_i] \right\},$$

while $(z^{0*} F^1)^*(z^{1*}) = \sum_{i=1}^n \left(z_i^{0*} \mathcal{T}_{\Omega_i, f_i}^{C_i}\right)^* (z_i^{1*})$, thus $(D_{h,\mathcal{T}}^S)$ becomes

$$(D_{h,\mathcal{T}}^S) \quad \sup_{\substack{\sum_{i=1}^n \lambda_i \leq 1, \ \lambda_i, z_i^{0*} \geq 0, \\ z_i^{1*} \in X^*, \ i=1,\ldots,n}} \left\{ -\sigma_S \left(-\sum_{i=1}^n z_i^{1*} \right) - \sum_{i=1}^n [(\lambda_i h_i)^*(z_i^{0*}) - \lambda_i a_i] \right.$$

$$\left. - \sum_{i=1}^n \left(z_i^{0*} \mathcal{T}_{\Omega_i, f_i}^{C_i}\right)^* (z_i^{1*}) \right\}.$$

In order to investigate further this dual problem, we separate in the sum $\sum_{i=1}^n (\lambda_i h_i)^*$ the terms with $\lambda_i > 0$ and the terms with $\lambda_i = 0$ as well as in $\sum_{i=1}^n \left(z_i^{0*} \mathcal{T}_{\Omega_i, f_i}^{C_i}\right)^*$ the terms with $z_i^{0*} > 0$ and the terms with $z_i^{0*} = 0$ in $(D_{h,\mathcal{T}}^S)$. Denote $I = \{i \in \{1, \ldots, n\} : z_i^{0*} > 0\}$ and $R = \{r \in \{1, \ldots, n\} : \lambda_r > 0\}$. If $i \in \{1, \ldots, n\} \setminus I$ it holds $(0 \cdot \mathcal{T}_{\Omega_i, f_i}^{C_i})^* = \sigma_X = \delta_{\{0_{X^*}\}}$, while when $i \in I$ one gets

$$\left(z_i^{0*} \mathcal{T}_{\Omega_i, f_i}^{C_i}\right)^* (z_i^{1*}) = \begin{cases} z_i^{0*} f_i^* \left(\frac{1}{z_i^{0*}} z_i^{1*}\right) + \sigma_{\Omega_i}(z_i^{1*}), & \text{if } \gamma_{C_i^0}(z_i^{1*}) \leq z_i^{0*}, \\ +\infty, & \text{otherwise.} \end{cases} \qquad (4.33)$$

Further, let us consider the case $r \in \{1, \ldots, n\} \setminus R$, i.e. $\lambda_r = 0$, then one has, since $z_r^{0*} \geq 0$,

$$(0 \cdot h_r)^*(z_r^{0*}) = \sup_{z_r \geq 0}\{z_r^{0*} z_r\} = \begin{cases} 0, & \text{if } z_r^{0*} = 0, \\ +\infty, & \text{otherwise.} \end{cases} \tag{4.34}$$

For $r \in R$, i.e. $\lambda_r > 0$, follows

$$(\lambda_r h_r)^*(z_r^{0*}) = \lambda_r h_r^*\left(\frac{z_r^{0*}}{\lambda_r}\right). \tag{4.35}$$

Hence, formula (4.34) implies that if $r \notin R$ then $z_r^{0*} = 0$, otherwise the values being not relevant for the dual problem, which means that $I \subseteq R$. Therefore $(D_{h,\mathcal{T}}^S)$ turns into

$$(D_{h,\mathcal{T}}^S) \quad \sup_{\substack{\lambda_i,\, z_i^{0*} \geq 0,\, i=1,\ldots,n, \\ I=\{i\in\{1,\ldots,n\}:z_i^{0*}>0\}\subseteq R=\{r\in\{1,\ldots,n\}:\lambda_r>0\}, \\ z_i^{1*}\in X^*,\, \gamma_{C_i^0}(z_i^{1*})\leq z_i^{0*},\, i\in I,\, \sum_{r\in R}\lambda_r\leq 1}} \left\{ -\sigma_S\left(-\sum_{i\in I} z_i^{1*}\right) - \sum_{r\in R}\lambda_r\left[h_r^*\left(\frac{z_r^{0*}}{\lambda_r}\right) - a_r\right] \right.$$

$$\left. -\sum_{i\in I}\left[z_i^{0*} f_i^*\left(\frac{1}{z_i^{0*}} z_i^{1*}\right) + \sigma_{\Omega_i}(z_i^{1*})\right]\right\}. \tag{4.36}$$

Remark 4.15. *Let $a_i = 0$, $i = 1, \ldots, n$. Taking the functions h_i as in Remark 4.14, their conjugates are $h_i^* = \delta_{(-\infty,1]}$, $i = 1, \ldots, n$, and the conjugate dual problem $(D_{\mathcal{T}}^S)$ to $(P_{\mathcal{T}}^S)$ reads as*

$$(D_{\mathcal{T}}^S) \quad \sup_{\substack{\lambda_i,\, z_i^{0*}\geq 0,\, i=1,\ldots,n,\, \sum_{r\in R}\lambda_r\leq 1, \\ I=\{i\in\{1,\ldots,n\}:z_i^{0*}>0\}\subseteq R=\{r\in\{1,\ldots,n\}:\lambda_r>0\}, \\ z_r^{0*}\leq\lambda_r,\, r\in R,\, z_i^{1*}\in X^*,\, \gamma_{C_i^0}(z_i^{1*})\leq z_i^{0*},\, i\in I}} \left\{ -\sigma_S\left(-\sum_{i\in I} z_i^{1*}\right) \right.$$

$$\left. -\sum_{i\in I}\left[z_i^{0*} f_i^*\left(\frac{1}{z_i^{0*}} z_i^{1*}\right) + \sigma_{\Omega_i}(z_i^{1*})\right]\right\}.$$

This dual problem can be simplified as follows.

Proposition 4.1. *The problem $(D_{\mathcal{T}}^S)$ can be equivalently written as*

$$(\widetilde{D}_{\mathcal{T}}^S) \quad \sup_{\substack{y_i^{0*}\geq 0,\, i=1,\ldots,n, \\ \widetilde{I}=\{i\in\{1,\ldots,n\}:y_i^{0*}>0\},\, y_i^{1*}\in X^*, \\ i\in\widetilde{I},\, \gamma_{C_i^0}(y_i^{1*})\leq y_i^{0*},\, i\in\widetilde{I},\, \sum_{i\in\widetilde{I}}y_i^{0*}\leq 1}} \left\{ -\sigma_S\left(-\sum_{i\in\widetilde{I}} y_i^{1*}\right) - \sum_{i\in\widetilde{I}}\left[y_i^{0*} f_i^*\left(\frac{1}{y_i^{0*}} y_i^{1*}\right) + \sigma_{\Omega_i}(y_i^{1*})\right]\right\}.$$

Proof. Take first a feasible element $(\lambda, z^{0*}, z^{1*}) = (\lambda_1, \ldots, \lambda_n, z_1^{0*}, \ldots, z_n^{0*}, z_I^{1*}) \in \mathbb{R}_+^n \times \mathbb{R}_+^n \times (X^*)^{|I|}$ to the problem $(D_{\mathcal{T}}^S)$, where by $z_I^{1*} \in (X^*)^{|I|}$ we denote the vector having as components z_i^{1*} with $i \in I$, and set $\widetilde{I} = I$, $y_i^{0*} = \lambda_i$, $i \in \widetilde{I}$, $y_j^{0*} = 0$, $j \notin \widetilde{I}$ and $y_i^{1*} = z_i^{1*}$, $i \in \widetilde{I}$, $y_j^{1*} = 0_{X^*}$, $j \notin \widetilde{I}$, then it follows from the feasibility of $(\lambda, z^{0*}, z^{1*})$ that $\sum_{i \in \widetilde{I}} y_i^{0*} \leq 1$, $y_i^{0*} > 0$, $y_i^{1*} \in X^*$, $\gamma_{C_i^0}(y_i^{1*}) \leq y_i^{0*}$, $i \in \widetilde{I}$ and $y_j^{0*} = 0$, $j \notin \widetilde{I}$, i.e. $(y^{0*}, y^{1*}) \in \mathbb{R}_+^n \times (X^*)^{|\widetilde{I}|}$ is feasible to the problem $(\widetilde{D}_{\mathcal{T}}^S)$. Hence, it holds $-\sigma_S\left(-\sum_{i \in I} z_i^{1*}\right) - \sum_{i \in I}\left[z_i^{0*} f_i^*((1/z_i^{0*}) z_i^{1*}) + \sigma_{\Omega_i}(z_i^{1*})\right] = -\sigma_S\left(-\sum_{i \in \widetilde{I}} y_i^{1*}\right) - \sum_{i \in \widetilde{I}}\left[y_i^{0*} f_i^*((1/y_i^{0*}) y_i^{1*}) + \sigma_{\Omega_i}(y_i^{1*})\right] \leq v(\widetilde{D}_{\mathcal{T}}^S)$ for all $(\lambda, z^{0*}, z^{1*})$ feasible to $(D_{\mathcal{T}}^S)$, i.e. $v(D_{\mathcal{T}}^S) \leq v(\widetilde{D}_{\mathcal{T}}^S)$.

To prove the opposite inequality, take a feasible element (y^{0*}, y^{1*}) of the problem $(\widetilde{D}_{\mathcal{T}}^S)$ and set $I = R = \widetilde{I}$, $z_i^{0*} = \lambda_i = y_i^{0*}$ and $z_i^{1*} = y_i^{1*}$ for $i \in I = R$ and $z_j^{0*} = \lambda_j = 0$ for $j \notin I = R$, then we have from the feasibility of (y^{0*}, y^{1*}) that $\sum_{r \in R} \lambda_r \leq 1$, $z_k^{1*} = \lambda_k > 0$, $k \in R$, $\lambda_l = 0$, $l \notin R$ and $\gamma_{C_i^0}(z_i^{1*}) \leq z_i^{1*}$, $i \in I$, which means that $(\lambda, z^{0*}, z^{1*})$ is a feasible element of $(D_{\mathcal{T}}^S)$ and it holds $-\sigma_S\left(-\sum_{i \in I} y_i^{1*}\right) - \sum_{i \in I}\left[y_i^{0*} f_i^*((1/y_i^{0*}) y_i^{1*}) + \sigma_{\Omega_i}(y_i^{1*})\right] = -\sigma_S\left(-\sum_{i \in I} z_i^{1*}\right) - \sum_{i \in I}\left[z_i^{0*} f_i^*((1/z_i^{0*}) z_i^{1*}) + \sigma_{\Omega_i}(z_i^{1*})\right] \leq v(D_{\mathcal{T}}^S)$ for all (y^{0*}, y^{1*}) feasible to $(\widetilde{D}_{\mathcal{T}}^S)$, which implies $v(\widetilde{D}_{\mathcal{T}}^S) \leq v(D_{\mathcal{T}}^S)$. Finally, it follows that $v(\widetilde{D}_{\mathcal{T}}^S) = v(D_{\mathcal{T}}^S)$. □

Also the general dual problem $(D_{h,\mathcal{T}}^S)$, it can be rewritten as follows.

Proposition 4.2. *The problem $(D_{h,\mathcal{T}}^S)$ can be equivalently written as*

$$(\widehat{D}_{h,\mathcal{T}}^S) \quad \sup_{\substack{\lambda_i,\, z_i^{0*} \geq 0,\, z_i^{1*} \in X^*,\, \sum_{i=1}^n \lambda_i \leq 1, \\ \gamma_{C_i^0}(z_i^{1*}) \leq z_i^{0*},\, i=1,\ldots,n}} \left\{ -\sigma_S\left(-\sum_{i=1}^n z_i^{1*}\right) - \sum_{i=1}^n [(\lambda_i h_i)^*\left(z_i^{0*}\right) \right.$$

$$\left. -\lambda_i a_i + \left(z_i^{0*} f_i\right)^*\left(z_i^{1*}\right) + \sigma_{\Omega_i}(z_i^{1*})] \right\}.$$

Proof. Let $(\lambda_1, \ldots, \lambda_n, z_1^{0*}, \ldots, z_n^{0*}, z_1^{1*}, \ldots, z_n^{1*})$ be a feasible solution to $(\widehat{D}_{h,\mathcal{T}}^S)$, then it follows from $r \notin R = \{r \in \{1, \ldots, n\} : \lambda_r > 0\}$ by (4.34) that $z_r^{0*} = 0$, i.e.

$$I = \left\{i \in \{1, \ldots, n\} : z_i^{0*} > 0\right\} \subseteq R,$$

and for $i \in \{1, \ldots, n\} \setminus I$ we have $0 \leq \gamma_{C_i^0}(z_i^{1*}) = \sigma_C(z_i^{1*}) \leq 0 \Leftrightarrow z_i^{1*} = 0_{X^*}$ (see [61, Proposition 2.2.3]). This means that $(\lambda_1, \ldots, \lambda_n, z_1^{1*}, \ldots, z_n^{1*}, z_I^{1*})$ is feasible to $(D_{h,\mathcal{T}}^S)$ and by (4.34) and (4.35) follows immediately that $v(\widehat{D}_{h,\mathcal{T}}^S) \leq v(D_{h,\mathcal{T}}^S)$.

Conversely, by the previous considerations it is clear that from any feasible solution to $(D_{h,\mathcal{T}}^S)$ one can immediately construct a feasible solution to $(\widehat{D}_{h,\mathcal{T}}^S)$ such that $v(D_{h,\mathcal{T}}^S) \leq v(\widehat{D}_{h,\mathcal{T}}^S)$ by taking $z_i^{1*} = 0_{X^*}$ for $i \in \{1, \ldots, n\} \setminus I$. □

Remark 4.16. *The index sets I and R of the dual problem $(D_{h,\mathcal{T}}^S)$ in (4.36) give a minute characterization of the set of feasible solutions and are useful in the further approach. From the numerical aspect however, they transform the dual (4.36) into a discrete optimization* *problem, making it very hard to solve. For this reason we use for theoretical approaches the dual $(D_{h,\mathcal{T}}^S)$ in the form of (4.36) and for numerical studies its equivalent dual formulation provided in Proposition 4.2. In this context, the dual $(\widetilde{D}_{\mathcal{T}}^S)$ is equivalent to*

$$(\widetilde{D}_{\mathcal{T}}^S) \qquad \sup_{\substack{z_i^{0*} \geq 0 \ ,z_i^{1*} \in X^*, \ \gamma_{C_i^0}(z_i^{1*}) \leq z_i^{0*}, \\ i=1,\ldots,n, \ \sum_{i=1}^n z_i^{0*} \leq 1}} \left\{ -\sigma_S\left(-\sum_{i=1}^n z_i^{1*} \right) - \sum_{i=1}^n \left[(z_i^{0*} f_i)^*(z_i^{1*}) + \sigma_{\Omega_i}(z_i^{1*}) \right] \right\}.$$

The weak duality for the primal-dual pair $(P_{h,\mathcal{T}}^S) - (D_{h,\mathcal{T}}^S)$ holds by construction, i.e. $v(P_{h,\mathcal{T}}^S) \geq v(D_{h,\mathcal{T}}^S)$, and we show that the considered hypotheses guarantee strong duality, too.

Theorem 4.3. *(strong duality) Between $(P_{h,\mathcal{T}}^S)$ and $(D_{h,\mathcal{T}}^S)$ strong duality holds, i.e. $v(P_{h,\mathcal{T}}^S) = v(D_{h,\mathcal{T}}^S)$ and the conjugate dual problem has an optimal solution*

$$(\overline{\lambda}_1, \ldots, \overline{\lambda}_n, \overline{z}_1^{0*}, \ldots, \overline{z}_n^{0*}, \overline{z}_{\overline{I}}^*) \in \mathbb{R}_+^n \times \mathbb{R}_+^n \times (X^*)^{|\overline{I}|} \qquad (4.37)$$

with the corresponding index sets $\overline{I} \subseteq \overline{R} \subseteq \{1, \ldots, n\}$.

Proof. The conclusion follows by Theorem 3.3, whose hypotheses are fulfilled as seen below. The properness and convexity properties of the involved functions and sets are guaranteed by the standing assumptions formulated in the beginning of the section. It remains to verify the fulfillment of a regularity condition. We use the generalized interior point regularity condition (RC_2^C) introduced in Section 3.2 for multi-composed optimization problems that is a variant of the one given in the general case in [106]. First, notice that f is lower semicontinuous, $K_0 = \mathbb{R}_+^n$ is closed and has a nonempty interior, S is closed, F^1 is $\mathbb{R}_+^n$-epi-closed, while the linear continuous function F^2 is obviously $\{0_{X^n}\}$-epi-closed. The continuity of F^2 voids (see Remark 3.6) the necessity of having int $K_1 \neq \emptyset$, a condition that is in this case not fulfilled. The other requirements of the regularity condition are fulfilled as well, namely $0_X \in \mathrm{sqri}((X \cap S) + X) = X$, $0_{\mathbb{R}^n} \in \mathrm{sqri}(F^1(\mathrm{dom}\,F^1) - \mathrm{dom}\,f + K_0) = \mathrm{sqri}(F^1(\mathrm{dom}\,F^1) - \mathbb{R}_+^n + \mathbb{R}_+^n) = \mathbb{R}^n$ and (recall that $\mathrm{dom}\,\mathcal{T}_{\Omega_i,f_i}^{C_i} = X, i = 1, \ldots, n$) $0_{X^n} \in \mathrm{sqri}(F^2(\mathrm{dom}\,F^2 \cap \mathrm{dom}\,g \cap S) - \mathrm{dom}\,F^1 + K_1) = \mathrm{sqri}(F^2(S) - X^n + \{0_{X^n}\}) = X^n.$ $\qquad \square$

The next statement is dedicated to deriving necessary and sufficient optimality conditions for the primal-dual pair $(P_{h,\mathcal{T}}^S) - (D_{h,\mathcal{T}}^S)$.

Theorem 4.4. *(optimality conditions) (a) Let $\overline{x} \in S$ be an optimal solution to the problem $(P_{h,\mathcal{T}}^S)$. Then there exists $(\overline{\lambda}_1, \ldots, \overline{\lambda}_n, \overline{z}_1^{0*}, \ldots, \overline{z}_n^{0*}, \overline{z}_{\overline{I}}^*) \in \mathbb{R}_+^n \times \mathbb{R}_+^n \times (X^*)^{|\overline{I}|}$ with the corresponding index sets $\overline{I} \subseteq \overline{R} \subseteq \{1, \ldots, n\}$ as an optimal solution to $(D_{h,\mathcal{T}}^S)$ such that*

$$(i) \ \max_{1 \leq j \leq n} \left\{ h_j\left(\mathcal{T}_{\Omega_j, f_j}^{C_j}(\overline{x}) \right) + a_j \right\} = \sum_{i \in \overline{I}} \overline{z}_i^{0*} \mathcal{T}_{\Omega_i, f_i}^{C_i}(\overline{x}) - \sum_{r \in \overline{R}} \overline{\lambda}_r \left[h_r^* \left(\frac{\overline{z}_r^{0*}}{\overline{\lambda}_r} \right) - a_r \right]$$

$$= \sum_{r \in \overline{R}} \overline{\lambda}_r \left[h_r\left(\mathcal{T}_{\Omega_r, f_r}^{C_r}(\overline{x}) \right) + a_r \right],$$

(ii) $\overline{\lambda}_r h_r^* \left(\frac{\overline{z}_r^{0*}}{\overline{\lambda}_r} \right) + \overline{\lambda}_r h_r \left(\mathcal{T}_{\Omega_r, f_r}^{C_r}(\overline{x}) \right) = \overline{z}_r^{0*} \mathcal{T}_{\Omega_r, f_r}^{C_r}(\overline{x}) \ \forall r \in \overline{R},$

(iii) $\overline{z}_i^{0*} \mathcal{T}_{\Omega_i, f_i}^{C_i}(\overline{x}) + \overline{z}_i^{0*} f_i^* \left(\frac{1}{\overline{z}_i^{0*}} \overline{z}_i^{1*} \right) + \sigma_{\Omega_i}(\overline{z}_i^{1*}) = \langle \overline{z}_i^{1*}, \overline{x} \rangle \ \forall i \in \overline{I},$

(iv) $\sum_{i \in \overline{I}} \langle \overline{z}_i^{1*}, \overline{x} \rangle = -\sigma_S \left(-\sum_{i \in \overline{I}} \overline{z}_i^{1*} \right),$

(v) $\max_{1 \le j \le n} \left\{ h_j \left(\mathcal{T}_{\Omega_j, f_j}^{C_j}(\overline{x}) \right) + a_j \right\} = h_r \left(\mathcal{T}_{\Omega_r, f_r}^{C_r}(\overline{x}) \right) + a_r \ \forall r \in \overline{R},$

(vi) $\sum_{r \in \overline{R}} \overline{\lambda}_r = 1, \ \overline{\lambda}_k > 0, \ k \in \overline{R}, \ \overline{\lambda}_l = 0, \ l \notin \overline{R}, \ \overline{z}_i^{0*} > 0, \ i \in \overline{I}, \ \text{and} \ \overline{z}_j^{0*} = 0, \ j \notin \overline{I},$

(vii) $\gamma_{C_i^0}(\overline{z}_i^{1*}) = \overline{z}_i^*, \ \overline{z}_i^{1*} \in X^* \setminus \{0_{X^*}\}, \ i \in \overline{I}, \ \text{and} \ \overline{z}_j^{1*} = 0_{X^*}, \ j \notin \overline{I}.$

(b) If there exists $\overline{x} \in S$ *such that for some* $(\overline{\lambda}_1, \ldots, \overline{\lambda}_n, \overline{z}_1^{0*}, \ldots, \overline{z}_n^{0*}, \overline{z}_{\overline{I}}^*) \in \mathbb{R}_+^n \times \mathbb{R}_+^n \times (X^*)^{|\overline{I}|}$ *with the corresponding index sets* $\overline{I} \subseteq \overline{R} \subseteq \{1, \ldots, n\}$ *the conditions* (i)-(vii) *are fulfilled, then* $\overline{x}$ *is an optimal solution to* $(P_{h,\mathcal{T}}^S)$, $(\overline{\lambda}_1, \ldots, \overline{\lambda}_n, \overline{z}_1^{0*}, \ldots, \overline{z}_n^{0*}, \overline{z}_{\overline{I}}^{1*})$ *is an optimal solution to* $(D_{h,\mathcal{T}}^S)$ *and* $v(P_{h,\mathcal{T}}^S) = v(D_{h,\mathcal{T}}^S).$

Proof. (a) By Theorem 3.4 we obtain the following necessary and sufficient optimality conditions for the primal-dual pair $(P_{h,\mathcal{T}}^S) - (D_{h,\mathcal{T}}^S)$

(i') $\max_{1 \le j \le n} \left\{ h_j \left(\mathcal{T}_{\Omega_j, f_j}^{C_j}(\overline{x}) \right) + a_j \right\} + \sum_{r \in \overline{R}} \overline{\lambda}_r \left[h_r^* \left(\frac{\overline{z}_r^{0*}}{\overline{\lambda}_r} \right) - a_r \right] = \sum_{i \in \overline{I}} \overline{z}_i^{0*} \mathcal{T}_{\Omega_i, f_i}^{C_i}(\overline{x}),$

(ii') $\sum_{i \in \overline{I}} \overline{z}_i^{0*} \mathcal{T}_{\Omega_i, f_i}^{C_i}(\overline{x}) + \sum_{i \in \overline{I}} \left[\overline{z}_i^{0*} f_i^* \left(\frac{1}{\overline{z}_i^{0*}} \overline{z}_i^{1*} \right) + \sigma_{\Omega_i}(\overline{z}_i^{1*}) \right] = \sum_{i \in \overline{I}} \langle \overline{z}_i^{1*}, \overline{x} \rangle,$

(iii') $\sum_{i \in \overline{I}} \langle \overline{z}_i^{1*}, \overline{x} \rangle + \sigma_S \left(-\sum_{i \in \overline{I}} \overline{z}_i^{1*} \right) = 0,$

(iv') $\sum_{r \in \overline{R}} \overline{\lambda}_r \le 1, \ \overline{\lambda}_k > 0, \ k \in \overline{R}, \ \overline{\lambda}_l = 0, \ l \notin \overline{R}, \ \overline{z}_i^{0*} > 0, \ i \in \overline{I}, \ \text{and} \ \overline{z}_j^{0*} = 0, \ j \notin \overline{I},$

(v') $\gamma_{C_i^0}(\overline{z}_i^{1*}) \le \overline{z}_i^{0*}, \ \overline{z}_i^{1*} \in X^*, \ i \in \overline{I}.$

Additionally, one has by Theorem 4.3 that $v(P_{h,a}^S) = v(D_{h,a}^S)$, i.e.

$$\max_{1 \le j \le n} \left\{ h_j \left(\mathcal{T}_{\Omega_j, f_j}^{C_j}(\overline{x}) \right) + a_j \right\} = -\sigma_S \left(-\sum_{i \in \overline{I}} \overline{z}_i^{1*} \right) - \sum_{r \in \overline{R}} \overline{\lambda}_r \left[h_r^* \left(\frac{\overline{z}_r^{0*}}{\overline{\lambda}_r} \right) - a_r \right]$$
$$- \sum_{i \in \overline{I}} \left[\overline{z}_i^{0*} f_i^* \left(\frac{1}{\overline{z}_i^{0*}} \overline{z}_i^{1*} \right) + \sigma_{\Omega_i}(\overline{z}_i^{1*}) \right],$$

that can be equivalently written as

$$
\left[\max_{1 \le j \le n} \left\{ h_j \left(\mathcal{T}^{C_j}_{\Omega_j, f_j}(\overline{x}) \right) + a_j \right\} - \sum_{r \in \overline{R}} \left(\overline{\lambda}_r h_r \left(\mathcal{T}^{C_r}_{\Omega_r, f_r}(\overline{x}) \right) + \overline{\lambda}_r a_r \right) \right]
$$

$$
+ \sum_{i \in \overline{I}} \left[\overline{z}_i^{0*} \left(\mathcal{T}^{C_i}_{\Omega_i, f_i}(\overline{x}) \right) + \overline{z}_i^{0*} f_i^* \left(\frac{1}{\overline{z}_i^{0*}} \overline{z}_i^{1*} \right) + \sigma_{\Omega_i}(\overline{z}_i^{1*}) - \langle \overline{z}_i^{1*}, \overline{x} \rangle \right]
$$

$$
+ \left[\sigma_S \left(- \sum_{i \in \overline{I}} \overline{z}_i^{1*} \right) + \sum_{i \in \overline{I}} \langle \overline{z}_i^{1*}, \overline{x} \rangle \right] + \sum_{i \in \overline{I}} \left[\overline{\lambda}_i h_i^* \left(\frac{\overline{z}_i^{0*}}{\overline{\lambda}_i} \right) + \overline{\lambda}_i h_i \left(\mathcal{T}^{C_i}_{\Omega_i, f_i}(\overline{x}) \right) - \overline{z}_i^{0*} \mathcal{T}^{C_i}_{\Omega_i, f_i}(\overline{x}) \right]
$$

$$
+ \sum_{r \in \overline{R} \setminus \overline{I}} \left[\overline{\lambda}_r h_r^*(0) + \overline{\lambda}_r h_r \left(\mathcal{T}^{C_r}_{\Omega_r, f_r}(\overline{x}) \right) - 0 \cdot \left(\mathcal{T}^{C_r}_{\Omega_r, f_r}(\overline{x}) \right) \right] = 0,
$$

where the last two sums arise from the fact that $\overline{I} \subseteq \overline{R}$. By Lemma 4.2 holds that the term within the first pair of brackets above is nonnegative. Moreover, by the Young-Fenchel inequality we have that the terms within the other brackets are nonnegative, too, and hence, it follows that all the terms within the brackets must be equal to zero. Combining the last statement with the optimality conditions $(i') - (v')$ yields (recall that $0 \le \gamma_{C_i^0}(z_i^{1*}) = \sigma_C(z_i^{1*}) \le 0 \Leftrightarrow z_i^{1*} = 0_{X^*}, i = 1, ..., n$)

(i) $\max\limits_{1 \le j \le n} \left\{ h_j \left(\mathcal{T}^{C_j}_{\Omega_j, f_j}(\overline{x}) \right) + a_j \right\} = \sum\limits_{i \in \overline{I}} \overline{z}_i^{0*} \mathcal{T}^{C_i}_{\Omega_i, f_i}(\overline{x}) - \sum\limits_{r \in \overline{R}} \overline{\lambda}_r \left[h_r^* \left(\frac{\overline{z}_r^{0*}}{\overline{\lambda}_r} \right) - a_r \right]$

$\quad = \sum\limits_{r \in \overline{R}} \left(\overline{\lambda}_r h_r \left(\mathcal{T}^{C_r}_{\Omega_r, f_r}(\overline{x}) \right) + \overline{\lambda}_r a_r \right),$

(ii) $\overline{\lambda}_r h_r^* \left(\frac{\overline{z}_r^{0*}}{\overline{\lambda}_r} \right) + \overline{\lambda}_r h_r \left(\mathcal{T}^{C_r}_{\Omega_r, f_r}(\overline{x}) \right) = \overline{z}_r^{0*} \mathcal{T}^{C_r}_{\Omega_r, f_r}(\overline{x}) \; \forall r \in \overline{R}$,

(iii) $\overline{z}_i^{0*} \mathcal{T}^{C_i}_{\Omega_i, f_i}(\overline{x}) + \overline{z}_i^{0*} f_i^* \left(\frac{1}{\overline{z}_i^{0*}} \overline{z}_i^{1*} \right) + \sigma_{\Omega_i}(\overline{z}_i^{1*}) = \langle \overline{z}_i^{1*}, \overline{x} \rangle \; \forall i \in \overline{I}$,

(iv) $\sum\limits_{i \in \overline{I}} \langle \overline{z}_i^{1*}, \overline{x} \rangle = -\sigma_S \left(- \sum\limits_{i \in \overline{I}} \overline{z}_i^{1*} \right)$,

(v) $\sum\limits_{r \in \overline{R}} \overline{\lambda}_r \le 1, \; \overline{\lambda}_k > 0, \; k \in \overline{R}, \; \overline{\lambda}_l = 0, \; l \notin \overline{R}, \; \overline{z}_i^{0*} > 0, \; i \in \overline{I}, \; \text{and} \; \overline{z}_j^{0*} = 0, \; j \notin \overline{I}$,

(vi) $\gamma_{C_i^0}(\overline{z}_i^{1*}) \le \overline{z}_i^{0*}, \; \overline{z}_i^{1*} \in X^*, \; i \in \overline{I}, \; \text{and} \; \overline{z}_j^{1*} = 0_{X^*}, \; j \notin \overline{I}$.

From conditions (i) and (v) we obtain that

$$\max_{1\le j\le n}\left\{h_j\left(\mathcal{T}^{C_j}_{\Omega_j,f_j}(\overline{x})\right)+a_j\right\}=\sum_{r\in\overline{R}}\left(\overline{\lambda}_r h_r\left(\mathcal{T}^{C_r}_{\Omega_r,f_r}(\overline{x})\right)+\overline{\lambda}_r a_r\right)$$

$$\le\sum_{r\in\overline{R}}\overline{\lambda}_r\max_{1\le j\le n}\left\{h_j\left(\mathcal{T}^{C_j}_{\Omega_j,f_j}(\overline{x})\right)+a_j\right\}\le\max_{1\le j\le n}\left\{h_j\left(\mathcal{T}^{C_j}_{\Omega_j,f_j}(\overline{x})\right)+a_j\right\},$$

which means on the one hand that

$$\sum_{r\in\overline{R}}\overline{\lambda}_r\max_{1\le j\le n}\left\{h_j\left(\mathcal{T}^{C_j}_{\Omega_j,f_j}(\overline{x})\right)+a_j\right\}=\max_{1\le j\le n}\left\{h_j\left(\mathcal{T}^{C_j}_{\Omega_j,f_j}(\overline{x})\right)+a_j\right\},$$

i.e. condition (v) can be written as

$$\sum_{r\in\overline{R}}\overline{\lambda}_r=1,\ \overline{\lambda}_k>0,\ k\in\overline{R},\ \overline{\lambda}_l=0,\ l\notin\overline{R},\ \overline{z}^{0*}_i>0,\ i\in\overline{I},\ \text{and}\ \overline{z}^{0*}_j=0,\ j\notin\overline{I},\quad(4.38)$$

and on the other hand that

$$\sum_{r\in\overline{R}}\left(\overline{\lambda}_r h_r\left(\mathcal{T}^{C_r}_{\Omega_r,f_r}(\overline{x})\right)+\overline{\lambda}_r a_r\right)=\sum_{r\in\overline{R}}\overline{\lambda}_r\max_{1\le j\le n}\left\{h_j\left(\mathcal{T}^{C_j}_{\Omega_j,f_j}(\overline{x})\right)+a_j\right\}$$

or, equivalently,

$$\sum_{r\in\overline{R}}\overline{\lambda}_r\left[\max_{1\le j\le n}\left\{h_j\left(\mathcal{T}^{C_j}_{\Omega_j,f_j}(\overline{x})\right)+a_j\right\}-h_r\left(\mathcal{T}^{C_r}_{\Omega_r,f_r}(\overline{x})\right)+a_r\right]=0.\quad(4.39)$$

As the brackets in (4.39) are nonnegative and $\overline{\lambda}_r>0$ for $r\in\overline{R}$, it follows that the terms inside the brackets must be equal to zero, more precisely,

$$\max_{1\le j\le n}\left\{h_j\left(\mathcal{T}^{C_j}_{\Omega_j,f_j}(\overline{x})\right)+a_j\right\}=h_r\left(\mathcal{T}^{C_r}_{\Omega_r,f_r}(\overline{x})\right)+a_r\ \forall r\in\overline{R}.\quad(4.40)$$

Further, Theorem 4.2 implies the existence of $\overline{p}_i,\ \overline{q}_i\in X$ such that

$$\mathcal{T}^{C_i}_{\Omega_i,f_i}(\overline{x})=\gamma_{C_i}(\overline{x}-\overline{p}_i-\overline{q}_i)+f_i(\overline{p}_i)+\delta_{\Omega_i}(\overline{q}_i)\ \forall i=1,..,n.$$

Employing the condition (iii) one gets

$$\overline{z}^{0*}_i\gamma_{C_i}(\overline{x}-\overline{p}_i-\overline{q}_i)-\overline{z}^{0*}_i f_i(\overline{p}_i)-\delta_{\Omega_i}(\overline{q}_i)+\overline{z}^{0*}_i f^*_i\left(\frac{1}{\overline{z}^{0*}_i}\overline{z}^{1*}_i\right)+\sigma_{\Omega_i}(\overline{z}^{1*}_i)=\langle\overline{z}^{1*}_i,\overline{x}\rangle,\quad(4.41)$$

equivalently writable as

$$\left[\overline{z}^{0*}_i\gamma_{C_i}(\overline{x}-\overline{p}_i-\overline{q}_i)-\langle\overline{z}^{1*}_i,\overline{x}-\overline{p}_i-\overline{q}_i\rangle\right]+\left[\overline{z}^{0*}_i f_i(\overline{p}_i)+\overline{z}^{0*}_i f^*_i\left(\frac{1}{\overline{z}^{0*}_i}\overline{z}^{1*}_i\right)-\langle\overline{z}^{1*}_i,\overline{p}_i\rangle\right]$$
$$+\left[\delta_{\Omega_i}(\overline{q}_i)+\sigma_{\Omega_i}(\overline{z}^{1*}_i)-\langle\overline{z}^{1*}_i,\overline{q}_i\rangle\right],\ i\in\overline{I}.$$
$$(4.42)$$

By the Young-Fenchel inequality all the brackets in (4.42) are nonnegative and must therefore be equal to zero, i.e.

$$\overline{z}_i^{0*}\gamma_{C_i}(\overline{x} - \overline{p}_i - \overline{q}_i) = \langle \overline{z}_i^{1*}, \overline{x} - \overline{p}_i - \overline{q}_i \rangle, \tag{4.43}$$

$$\overline{z}_i^{0*}f_i(\overline{p}_i) + \overline{z}_i^{0*}f_i^*\left(\frac{1}{\overline{z}_i^{0*}}\overline{z}_i^{1*}\right) = \langle \overline{z}_i^{1*}, \overline{p}_i \rangle,$$

$$\delta_{\Omega_i}(\overline{q}_i) + \sigma_{\Omega_i}(\overline{z}_i^{1*}) = \langle \overline{z}_i^{1*}, \overline{q}_i \rangle, \ i \in \overline{I}.$$

Now, by (4.43), condition (vi) and Lemma 4.3 (the generalized Cauchy-Schwarz inequality) yield

$$\overline{z}_i^{0*}\gamma_{C_i}(\overline{x} - \overline{p}_i - \overline{q}_i) = \langle \overline{z}_i^{1*}, \overline{x} - \overline{p}_i - \overline{q}_i \rangle \leq \gamma_{C_i^0}(\overline{z}_i^{1*})\gamma_{C_i}(\overline{x} - \overline{p}_i - \overline{q}_i) \leq \overline{z}_i^{0*}\gamma_{C_i}(\overline{x} - \overline{p}_i - \overline{q}_i), \tag{4.44}$$

which means that condition (vi) can be expressed as

$$\gamma_{C_i^0}(\overline{z}_i^{1*}) = \overline{z}_i^{0*}, \ \overline{z}_i^{1*} \in X^* \setminus \{0_{X^*}\}, \ i \in \overline{I}, \text{ and } \overline{z}_j^{1*} = 0_{X^*}, \ j \notin \overline{I}. \tag{4.45}$$

The optimality conditions (i)-(vi), (4.38), (4.40) and (4.45) deliver the desired statement.

(b) All the calculations done in (a) can also be made in the reverse order, yielding thus the conclusion. $\qquad \square$

Remark 4.17. *If we consider the situation when the set-up costs are arbitrary, i.e. a_i can also be negative, $i = 1, \ldots, n$, then the conjugate function of f looks like (see Remark 4.2)*

$$f^*(z_1^*, \ldots, z_n^*) = \min_{\substack{\sum_{i=1}^n \lambda_i = 1, \ \lambda_i \geq 0, \\ i=1,\ldots,n}} \left\{ \sum_{i=1}^n [(\lambda_i h_i)^*(z_i^{0*}) - \lambda_i a_i] \right\}.$$

As a consequence, the corresponding dual problem turns out to be almost the same one as in (4.36) with the additional constraint $\sum_{r \in R} \lambda_r = 1$ and all the statements given in this subsection can be easily adapted for this general case where the set-up costs are arbitrary.

4.2.1 Special case I

We study now the location problem involved in the economical scenario discussed in Remark 4.13 (we set $C_i = -C_i$), i.e.

$$(P_{\gamma G}, \mathcal{T}) \qquad \inf_{x \in X} \max_{1 \leq i \leq n} \left\{ \mathcal{T}_{\Omega_i, \gamma_{G_i}}^{C_i}(x) \right\},$$

and its dual problem (cf. Proposition 4.1, note that $S = X$, $a_i = 0$ and $f_i^* = \delta_{G_i^0}$, $i = 1, \ldots, n$)

$$(D_{\gamma G}, \mathcal{T}) \qquad \sup_{\substack{z_i^{0*} \geq 0, \ i=1,\ldots,n, \ I=\{i \in \{1,\ldots,n\}: z_i^{0*} > 0\}, \\ z_i^{1*} \in X^*, \ i \in I, \ \gamma_{C_i^0}(z_i^{1*}) \leq z_i^{0*}, \\ \gamma_{G_i^0}(z_i^{1*}) \leq z_i^{0*}, i \in I, \ \sum_{i \in I} z_i^{0*} \leq 1, \ \sum_{i \in I} z_i^{1*} = 0_{X^*}}} \left\{ -\sum_{i \in I} \sigma_{\Omega_i}\left(z_i^{1*}\right) \right\}.$$

Theorem 4.3 yields the following duality statement for the primal-dual pair $(P_{\gamma G}, \tau)$-$(D_{\gamma G}, \tau)$.

Theorem 4.5. *(strong duality) Between $(P_{\gamma G}, \tau)$ and $(D_{\gamma G}, \tau)$ holds strong duality, i.e.* $v(P_{\gamma G}, \tau) = v(D_{\gamma G}, \tau)$ *and the dual problem has an optimal solution.*

The necessary and sufficient optimality conditions for the primal-dual pair of optimization problems $(P_{\gamma G}, \tau) - (D_{\gamma G}, \tau)$ can be obtained by using the same ideas as in Theorem 4.4.

Theorem 4.6. *(optimality conditions) (a) Let $\overline{x} \in X$ be an optimal solution to the problem* $(P_{\gamma G}, \tau)$. *Then there exists an optimal solution to $(D_{\gamma G}, \tau)$ $(\overline{z}_1^{0*}, \ldots, \overline{z}_n^{0*}, \overline{z}_1^{1*}, \ldots, \overline{z}_n^{1*})$ with the corresponding index set $\overline{I} \subseteq \{1, \ldots, n\}$ such that*

(i) $\max\limits_{1 \le j \le n} \left\{ \mathcal{T}_{\Omega_j, \gamma_{G_j}}^{C_j}(\overline{x}) \right\} = \sum\limits_{i \in \overline{I}} \overline{z}_i^{0*} \mathcal{T}_{\Omega_i, \gamma_{G_i}}^{C_i}(\overline{x})$,

(ii) $\overline{z}_i^{0*} \mathcal{T}_{\Omega_i, \gamma_{G_i}}^{C_i}(\overline{x}) + \sigma_{\Omega_i}(\overline{z}_i^{1*}) = \langle \overline{z}_i^{1*}, \overline{x} \rangle \; \forall i \in \overline{I}$,

(iii) $\sum\limits_{i \in \overline{I}} \overline{z}_i^{1*} = 0_{X^*}$,

(iv) $\max\limits_{1 \le j \le n} \left\{ \mathcal{T}_{\Omega_j, \gamma_{G_j}}^{C_j}(\overline{x}) \right\} = \mathcal{T}_{\Omega_i, \gamma_{G_i}}^{C_i}(\overline{x}) \; \forall i \in \overline{I}$,

(v) $\sum\limits_{i \in \overline{I}} \overline{z}_i^{0*} = 1$, $\overline{z}_i^{0*} > 0$, $i \in \overline{I}$, *and* $\overline{z}_j^{0*} = 0$, $j \notin \overline{I}$,

(vi) $\gamma_{C_i^0}(\overline{z}_i^{1*}) = \overline{z}_i^{0*}$, $\overline{z}_i^{1*} \in X^* \setminus \{0_{X^*}\}$, $\gamma_{G_i^0}(\overline{z}_i^{1*}) \le \gamma_{C_i^0}(\overline{z}_i^{1*})$, $i \in \overline{I}$.

(b) If there exists $\overline{x} \in \mathcal{H}$ such that for some $(\overline{z}_1^{0*}, \ldots, \overline{z}_n^{0*}, \overline{z}_1^{1*}, \ldots, \overline{z}_n^{1*})$ and the corresponding index set $\overline{I}$ the conditions (i)-(vi) are fulfilled, then $\overline{x}$ is an optimal solution to $(P_{\gamma G}, \tau)$, $(\overline{z}_1^{0*}, \ldots, \overline{z}_n^{0*}, \overline{z}_1^{1*}, \ldots, \overline{z}_n^{1*})$ is an optimal solution to $(D_{\gamma G}, \tau)$ and $v(P_{\gamma G}, \tau) = v(D_{\gamma G}, \tau)$.

Proof. As $h_i^* = \delta_{(-\infty, 1]}$ for all $i = 1, \ldots, n$, one has from the optimality condition (ii) of Theorem 4.4 that $\overline{z}_r^{0*} \mathcal{T}_{\Omega_r, \gamma_{G_r}}^{C_r}(\overline{x}) = \overline{\lambda}_r^* \mathcal{T}_{\Omega_r, \gamma_{G_r}}^{C_r}(\overline{x})$ for all $r \in \overline{R}$, which in turn yields that $\overline{I} = \overline{R}$ and $\overline{\lambda}_i = \overline{z}_i^{0*}$ for all $i \in \overline{I}$ (as $0 < z_r^{0*} \le \lambda_r$ and $\overline{I} \subseteq \overline{R}$). Furthermore, as $f_i^* = \delta_{G_i^0}$, it follows by the optimality conditions (iii) and (vii) of Theorem 4.4 that $\gamma_{G_i^0}(\overline{z}_i^{1*}) \le \gamma_{C_i^0}(\overline{z}_i^{1*})$ for all $i \in \overline{I}$. Summing up these facts with the optimality conditions of Theorem 4.4 yields the desired statement. $\qquad \square$

We use the optimality conditions listed in Theorem 4.6 to provide a more exact characterization to the optimal solutions to the optimization problem $(P_{\gamma G}, \tau)$.

Theorem 4.7. *Let $\cap_{i \in \overline{I}} \Omega_i = \emptyset$, $0 \in \operatorname{int} G_i$, $C_i^0 \cap G_i \cap \operatorname{dom} \sigma_{\Omega_i} \ne \emptyset$ for all $i \in \overline{I}$, and $\overline{x} \in X$ be an optimal solution to the optimization problem $(P_{\gamma G}, \tau)$. If $(\overline{z}_1^{0*}, \ldots, \overline{z}_n^{0*}, \overline{z}_1^{1*}, \ldots, \overline{z}_n^{1*}) \in \mathbb{R}_+^n \times (X^*)^n$ is an optimal solution to $(D_{\gamma G}, \tau)$ with the corresponding $\overline{I} \subseteq \{1, \ldots, n\}$, then*

$$\overline{x} \in \bigcap_{i \in \overline{I}} \left[\partial \left(v(D_{\gamma G}, \tau) \gamma_{C_i^0} \right)(\overline{z}_i^{1*}) + \partial \sigma_{\Omega_i}(\overline{z}_i^{1*}) \right].$$

Proof. From $0_X \in \operatorname{int} C_i$ and $0_X \in \operatorname{int} G_i$ follows that $0_{X^*} \in \operatorname{int} C_i^0$ and $0_{X^*} \in \operatorname{int} G_i^0$ such that $\gamma_{C_i^0}$ and $\gamma_{G_i^0}$ are continuous for all $i \in \overline{I}$. Hence, Theorem 4.2 secures the existence of $\phi_i \in X$ and $\psi_i \in \Omega_i$ such that $\mathcal{T}_{\Omega_i,\gamma_{G_i}}^{C_i}(\overline{x}) = \gamma_{C_i}(\overline{x} - \phi_i - \psi_i) + \gamma_{G_i}(\phi_i)$, $i \in \overline{I}$. Further, we have by the optimality conditions (ii) and (iv) of Theorem 4.6

$$(\gamma_{C_i}(\overline{x} - \phi_i - \psi_i) + \gamma_{G_i}(\phi_i))\gamma_{C_i^0}(\overline{z}_i^{1*}) + \sigma_{\Omega_i}(\overline{z}_i^{1*}) = \langle \overline{z}_i^{1*}, \overline{x} \rangle$$

$$\Leftrightarrow \left[\gamma_{C_i}(\overline{x} - \phi_i - \psi_i)\gamma_{C_i^0}(\overline{z}_i^{1*}) - \langle \overline{z}_i^{1*}, \overline{x} - \phi_i - \psi_i \rangle \right] + \left[\gamma_{G_i}(\phi_i)\gamma_{C_i^0}(\overline{z}_i^{1*}) - \langle \overline{z}_i^{1*}, \phi_i \rangle \right]$$

$$+ \left[\sigma_{\Omega_i}(\overline{z}_i^{1*}) - \langle \overline{z}_i^{1*}, \psi_i \rangle \right] = 0, \ i \in \overline{I},$$

from which follows with the Young-Fenchel inequality that

$$\overline{x} - \phi_i - \psi_i \in \partial \left(\gamma_{C_i}(\overline{x} - \phi_i - \psi_i)\gamma_{C_i^0} \right)(\overline{z}_i^{1*}), \tag{4.46}$$

$$\phi_i \in \partial \left(\gamma_{G_i}(\phi_i)\gamma_{C_i^0} \right)(\overline{z}_i^{1*}), \tag{4.47}$$

$$\psi_i \in \partial \sigma_{\Omega_i}(\overline{z}_i^{1*}), \ i \in \overline{I}. \tag{4.48}$$

If $\gamma_{C_i}(\overline{x} - \phi_i - \psi_i) + \gamma_{G_i}(\phi_i) = 0$, $i \in \overline{I}$, then we have by (4.46), (4.47) and (4.48) that $\overline{x} \in \partial \sigma_{\Omega_i}(\overline{z}_i^{1*})$ for all $i \in \overline{I}$, such that $\overline{x} \in \Omega_i$ for all $i \in \overline{I}$, which contradicts our assumption. If there exists $i \in \overline{I}$ such that $\gamma_{C_i}(\overline{x} - \phi_i - \psi_i) = 0$, then $v(D_{\gamma_G}, \tau) = \gamma_{G_i}(\phi_i) > 0$ and we get by (4.46), (4.47) and (4.48) that

$$\overline{x} - \psi_i \in \partial \delta_{X^*}(\overline{z}_i^{1*}) + \partial \left(\gamma_{G_i}(\phi_i)\gamma_{C_i^0} \right)(\overline{z}_i^{1*}) = \{0_{X^*}\} + \gamma_{G_i}(\phi_i)\partial \gamma_{C_i^0}(\overline{z}_i^{1*})$$

$$= v(D_{\gamma_G}, \tau)\partial \gamma_{C_i^0}(\overline{z}_i^{1*}).$$

If there exists $i \in \overline{I}$ such that $\gamma_{C_i}(\overline{x} - \phi_i - \psi_i) > 0$ and $\gamma_{G_i}(\phi_i) = 0$, then it follows in a similar way by (4.46), (4.47) and (4.48) that

$$\overline{x} - \psi_i \in v(D_{\gamma_G}, \tau)\partial \gamma_{C_i^0}(\overline{z}_i^{1*}).$$

Finally, if there exists $i \in \overline{I}$ such that $\gamma_{C_i}(\overline{x} - \phi_i - \psi_i) > 0$ and $\gamma_{G_i}(\phi_i) > 0$, then one has by (4.46), (4.47) and (4.48) that that

$$\overline{x} - \psi_i \in \partial \left((\gamma_{C_i}(\overline{x} - \phi_i - \psi_i) + \gamma_{G_i}(\phi_i))\gamma_{C_i^0} \right)(\overline{z}_i^{1*}) = v(D_{\gamma_G}, \tau)\partial \gamma_{C_i^0}(\overline{z}_i^{1*}).$$

In summary, we have $\overline{x} - \psi \in v(D_{\gamma_G}, \tau)\partial \gamma_{C_i^0}(\overline{z}_i^{1*})$, which implies that

$$\overline{x} \in v(D_{\gamma_G}, \tau)\partial \gamma_{C_i^0}(\overline{z}_i^{1*}) + \partial \sigma_{\Omega_i}(\overline{z}_i^{1*}) \forall i \in \overline{I}.$$

$\square$

Remark 4.18. *Let $\mathcal{H}$ be a real Hilbert space, $\beta_i > 0$, $p_i \in \mathcal{H}$, $C_i = \{x \in \mathcal{H} : \beta_i \|x\|_{\mathcal{H}} \leq 1\}$, γ_{G_i} a norm and $\Omega_i = \{p_i\}$, $i = 1, \ldots, n$, with $p_1, \ldots, p_n$ distinct, then one has by Theorem 4.7*

$$\overline{x} = \frac{v(D_{\gamma_G}, \tau)}{\beta_i \|\overline{z}_i^{1*}\|_{\mathcal{H}}} \overline{z}_i^{1*} + p_i \ \forall i \in \overline{I}.$$

Note that if $v(P_{\gamma_G}, \tau) = 0$, then $\gamma_{C_i}(\overline{x} - p_i - \overline{z}_i) + \gamma_{G_i}(\overline{z}_i) = 0$ for all $i = 1, \ldots, n$, which means that $z_i = 0_{\mathcal{H}}$ and $\overline{x} = p_i$ for all $i = 1, \ldots, n$, i.e. one gets a contradiction. Therefore, taking into consideration that $v(P_{\gamma_G}, \tau) > 0$ and the strong duality statement, one gets $v(D_{\gamma_G}, \tau) > 0$ and by the optimality condition (iii) of Theorem 4.6 follows

$$\sum_{i \in \overline{I}} \frac{\beta_i \|\overline{z}_i^{1*}\|_{\mathcal{H}}}{v(D_{\gamma_G}, \tau)} (\overline{x} - p_i) = \sum_{i \in \overline{I}} \overline{z}_i^{1*} = 0_{\mathcal{H}} \Leftrightarrow \overline{x} = \frac{1}{\sum_{i \in \overline{I}} \beta_i \|\overline{z}_i^{1*}\|_{\mathcal{H}}} \sum_{i \in \overline{I}} \beta_i \|\overline{z}_i^{1*}\|_{\mathcal{H}} p_i.$$

Remark 4.19. *Let $\cap_{i \in \overline{I}} \Omega_i = \emptyset$, $0 \in \mathrm{int}\, G_i$ and $\gamma_{C_i^0}(x) = \gamma_{G_i^0}(x) = 0$ if and only if $x = 0_X$, $i = 1, \ldots, n$.*

(i) Following Proposition 4.2, the dual problem $(\widetilde{D}_{\gamma_G}, \tau)$ can be rewritten as

$$(\widetilde{D}_{\gamma_G}, \tau) \quad \sup_{\substack{y_i^* \in X^*,\ i=1,\ldots,n,\ \sum_{i=1}^n y_i^* = 0_{X^*}, \\ \sum_{i=1}^n \max\{\gamma_{C_i^0}(y_i^*),\ \gamma_{G_i^0}(y_i^*)\} \leq 1}} \left\{ -\sum_{i=1}^n \sigma_{\Omega_i}(y_i^*) \right\},$$

consequently $v(P_{\gamma_G}, \tau) = v(\widetilde{D}_{\gamma_G}, \tau)$.

(ii) As the Slater constraint qualification corresponding to $(\widetilde{D}_{\gamma_G}, \tau)$ is fulfilled (for instance for $y_i^* = 0_{X^*}$, $i = 1, \ldots, n$), there holds strong duality for it and its Lagrange dual problem, that can be reduced after some calculations to

$$(D\widetilde{D}_{\gamma_G}, \tau) \quad \inf_{\lambda \geq 0,\ x \in X} \left\{ \lambda + \sum_{i=1}^n \sup_{y_i^* \in X^*} \left\{ \langle x, y_i^* \rangle - \lambda \max\left\{ \gamma_{C_i^0}(y_i^*),\ \gamma_{G_i^0}(y_i^*) \right\} - \sigma_{\Omega_i}(y_i^*) \right\} \right\}. \tag{4.49}$$

Since $\lambda = 0$ implies, taking into consideration that $\cap_{i=1}^n \Omega_i = \emptyset$, that the value of the objective function of $(D\widetilde{D}_{\gamma_G}, \tau)$ is $+\infty$, one can write $\lambda > 0$ in the constraints of $(D\widetilde{D}_{\gamma_G}, \tau)$. Moreover, since $0_{X^*} \in \mathrm{dom}\, \gamma_{C_i^0} \cap \mathrm{dom}\, \gamma_{G_i^0} \cap \mathrm{dom}\, \sigma_{\Omega_i}$ and σ_{Ω_i} is continuous for all $i = 1, \ldots, n$, [14, Theorem 3.5.8.(a)] yields

$$\sup_{y_i^* \in X^*} \left\{ \langle x, y_i^* \rangle - \lambda \max\left\{ \gamma_{C_i^0}(y_i^*),\ \gamma_{G_i^0}(y_i^*) \right\} - \sigma_{\Omega_i}(y_i^*) \right\}$$

$$= \min_{y_i \in \Omega} \left\{ \lambda \max\left\{ \gamma_{C_i^0}(\cdot),\ \gamma_{G_i^0}(\cdot) \right\}^* \left(\frac{1}{\lambda}(x - y_i) \right) \right\}. \tag{4.50}$$

(iii) For any $i = 1, \ldots, n$, the conjugate of $\max\left\{\gamma_{C_i^0}(\cdot),\ \gamma_{G_i^0}(\cdot)\right\}$ from (4.50) becomes

$$\max\left\{\gamma_{C_i^0}(\cdot),\ \gamma_{G_i^0}(\cdot)\right\}^*(x) = \sup_{\substack{x^* \in X^*,\ t \geq 0, \\ \gamma_{C_i^0}(x^*) \leq t,\ \gamma_{G_i^0}(x^*) \leq t}} \{\langle x, x^*\rangle - t\},\ i = 1, \ldots, n. \tag{4.51}$$

As the Slater constraint qualification for the problem in the right-hand side of (4.51) is obviously fulfilled, one obtains via strong Lagrange duality

$$\max\left\{\gamma_{C_i^0}(\cdot),\ \gamma_{G_i^0}(\cdot)\right\}^*(x) = \min_{\substack{\alpha \geq 0,\ \beta \geq 0, \\ \alpha+\beta \leq 1}} \left(\alpha\gamma_{C_i^0} + \beta\gamma_{G_i^0}\right)^*(x). \tag{4.52}$$

Note that a more general formula for this conjugate can be found in [27]. Recall that $0_X \in \operatorname{int} C_i$ and $0_X \in \operatorname{int} G_i$, which implies that $0_{X^*} \in \operatorname{int} C_i^0$ and $0_{X^*} \in \operatorname{int} G_i^0$ and thus, $\operatorname{dom}\gamma_{C_i^0} = \operatorname{dom}\gamma_{G_i^0} = X^*$. Hence, we have $0 \cdot \gamma_{C_i^0} = 0 \cdot \gamma_{G_i^0} = \delta_{X^*}$. We apply [14, Theorem 3.5.8.(a)] to the formula in the right-hand side of (4.52), where the minimum is assumed to be attained at $(\bar{\alpha}, \bar{\beta})$.

If $\bar{\alpha} = 0$ and $\bar{\beta} > 0$, then we have

$$\min_{0 \leq \beta \leq 1} \left(\delta_{X^*} + \beta\gamma_{G_i^0}\right)^*(x) = \min_{0 < \beta \leq 1}\left\{\delta_{\{0_X\}}(x - z_i) + \beta\gamma_{G_i^0}^*\left(\frac{1}{\beta}z_i\right)\right\}$$
$$= \begin{cases} 0, & \text{if } 0 < \bar{\beta} \leq 1,\ \gamma_{G_i}(x) \leq \bar{\beta}, \\ +\infty, & \text{otherwise} \end{cases} = \begin{cases} 0, & \text{if } \gamma_{G_i}(x) \leq 1 \\ +\infty, & \text{otherwise.} \end{cases} \tag{4.53}$$

If $\bar{\alpha} > 0$ and $\bar{\beta} = 0$, then one gets similarly that

$$\min_{0 \leq \alpha \leq 1} \left(\alpha\gamma_{C_i^0} + \delta_{X^*}\right)^*(x) = \begin{cases} 0, & \text{if } \gamma_{C_i}(x) \leq 1 \\ +\infty, & \text{otherwise.} \end{cases} \tag{4.54}$$

Finally, when $\bar{\alpha} > 0$ and $\bar{\beta} > 0$, then

$$\min_{\substack{\alpha \geq 0,\ \beta \geq 0, \\ \alpha+\beta \leq 1}} \left(\alpha\gamma_{C_i^0} + \beta\gamma_{G_i^0}\right)^*(x) = \min_{z_i \in X}\left\{\bar{\alpha}\gamma_{C_i^0}^*\left(\frac{1}{\bar{\alpha}}(x - z_i)\right) + \bar{\beta}\gamma_{G_i^0}^*\left(\frac{1}{\bar{\beta}}z_i\right)\right\}$$
$$= \begin{cases} 0, & \text{if } \gamma_{C_i}(x - z_i) + \gamma_{G_i}(z_i) \leq 1, \\ +\infty, & \text{otherwise.} \end{cases} \tag{4.55}$$

As $\gamma_{C_i^0}(x) = \gamma_{C_i^0}(x) = 0 \Leftrightarrow x = 0_X$, it follows from (4.53), (4.54) and (4.55) that

$$\min_{\substack{\alpha \geq 0,\ \beta \geq 0, \\ \alpha+\beta \leq 1}} \left(\alpha\gamma_{C_i^0} + \beta\gamma_{G_i^0}\right)^*(x) = \begin{cases} 0, & \text{if } \gamma_{C_i}(x - z_i) + \gamma_{G_i}(z_i) \leq 1, \\ +\infty, & \text{otherwise,} \end{cases}\ i = 1, \ldots, n. \tag{4.56}$$

(iv) Bringing (4.49), (4.50) and (4.56) together allows us consecutively to reformulate the Lagrange dual problem $(D\widetilde{D}_{\gamma_G},\ \tau)$ as

$$\min_{\substack{\lambda > 0,\ x \in X,\ y_i \in \Omega_i,\ z_i \in X, \\ \gamma_{C_i}(x - y_i - z_i) + \gamma_{G_i}(z_i) \leq \lambda,\ i=1,\ldots,n}} \lambda = \min_{\substack{\lambda > 0,\ x \in X, \\ \min\limits_{y_i \in \Omega_i,\ z_i \in X}\left\{\gamma_{C_i}(x-y_i-z_i)+\gamma_{G_i}(z_i)\right\} \leq \lambda,\ i=1,\ldots,n}} \lambda$$

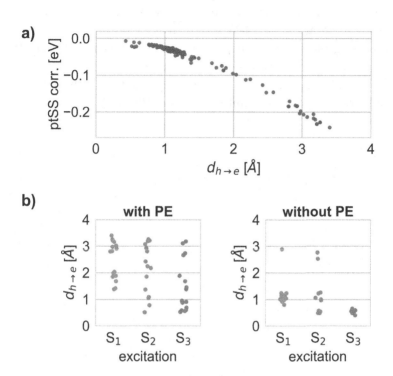

Figure 6.8: Analysis of Lf-W36 CT states using electron-hole distances. a) Relationship between ptSS correction and electron-hole distance $d_{h \to e}$ for S_1 and S_2. The electron-hole distance quantifies a permanent CT, which is also reflected by a large shift in electron density, leading to a larger ptSS correction term. b) Effect of PE on electron-hole distances. In isolation, i.e., without the protein environment included through PE, the character of the excitation is shifted from CT to LE.[1]

essential in mediating the CT excitation in order to efficiently protect flavin derivatives from photodegradation.

The calculations with PE-ADC(2) clearly show that the protein and solvent environment is capable of stabilizing and thus promoting the CT excitation on Lf-W36 which is essential to the current understanding of the dodecin photocycle. In experimental studies, the photocycle was manipulated through W36 mutants with different redox potentials[122]. Similarly, one could computationally probe and design future active site mutants guided by PE-ADC.

7 Conclusions and Outlook

In the presented thesis, the first combination of the PE model with ADC was shown, including a derivation of the theoretical methodology, an implementation ready for productive calculations, and applications to show the power of the joint method. A novel library providing a flexible framework for PE calculations, CPPE, was implemented and interfaced with the popular *Q-Chem* program package. The excitation energies in the PE-ADC approach are corrected by a state-specific (ptSS) and linear-response-type (ptLR) term. Test calculations were presented for water-coordinated PNA showing that the presented approach performs well when compared to super-system ADC calculations. Also, a comparative study of PE-ADC and PCM-ADC on lumiflavin showed that PE-ADC is well-behaved when modeling bulk solvation effects. The case study using PE-ADC to investigate the photoprotection mechanism in archaeal dodecin shed light on the role of the protein and solvent environment: Dodecin is able to stabilize the low-lying CT state, enabling an efficient excited state quenching mechanism. To analyze such environment contributions, state-of-the-art wave function and excited state analysis methods were empolyed. As a matter of fact, the inclusion of the protein environment did not dramatically increase the computation time compared to canonical ADC. Particularly, the computational cost of the ADC calculation as such is literally the same as without PE. In future applications, the PE model should also be combined with more ADC variants, such as the resolution-of-the-identity (RI) ADC[123,124] or core-valence-separation (CVS) ADC[37,38]. With CVS-ADC, it might be possible to calculate accurate X-ray emission spectra in condensed phase[125,126]. Once the ISR is available for ionized states with ionization-potential (IP) ADC, one will also be able to combine it with the PE model. Such

© The Editor(s) (if applicable) and The Author(s), under exclusive license to
Springer Fachmedien Wiesbaden GmbH, part of Springer Nature 2020
M. Scheurer, *Polarizable Embedding for the Algebraic- Diagrammatic
Construction Scheme*, BestMasters, https://doi.org/10.1007/978-3-658-31281-7_7

calculations could be of great importance to study, e.g., ionization processes in DNA[127]. In contrast to the here presented *a posteriori* corrections, it could be of importance to include environmental relaxation in the ADC Hamiltonian for ionized states because the induced dipole moments most likely undergo a large change upon ionization. Ongoing projects using PE-ADC try to examine electronic excitations in flavodoxin, another flavin-binding protein[128], and channelrhodopsin[30]. Of note, NMR properties of embedded molecules are to be implemented as well using PE with orbital-optimized MP2 theory[129] in finite-difference calculations.

In summary, PE-ADC can be employed to study a plethora of biochemically relevant systems in order to obtain highly accurate excited state descriptions in complex, polarizable environments at reasonable computational cost. The presented thesis provides the necessary and comprehensive cornerstone for this novel method and its future applicability.

Notes

1. Reprinted with permission from "Polarizable Embedding Combined with the Algebraic Diagrammatic Construction: Tackling Excited States in Biomolecular Systems; Maximilian Scheurer, Michael F. Herbst, Peter Reinholdt, Jógvan Magnus Haugaard Olsen, Andreas Dreuw, and Jacob Kongsted. Journal of Chemical Theory and Computation 2018 14 (9), 4870-4883". Copyright 2018 American Chemical Society.

Bibliography

[1] Mohseni, M.; Omar, Y.; Engel, G. S.; Plenio, M. B. *Quantum Effects in Biology*; 2014; pp 1–399.

[2] Senn, H. M.; Thiel, W. QM/MM Methods for Biomolecular Systems. *Angew. Chemie Int. Ed.* **2009**, *48*, 1198–1229.

[3] Wallrapp, F. H.; Guallar, V. Mixed quantum mechanics and molecular mechanics methods: Looking inside proteins. *Wiley Interdiscip. Rev. Comput. Mol. Sci.* **2011**, *1*, 315–322.

[4] Van Der Kamp, M. W.; Mulholland, A. J. Combined quantum mechanics/molecular mechanics (QM/MM) methods in computational enzymology. *Biochemistry* **2013**, *52*, 2708–2728.

[5] Riplinger, C.; Sandhoefer, B.; Hansen, A.; Neese, F. Natural triple excitations in local coupled cluster calculations with pair natural orbitals. *J. Chem. Phys.* **2013**, *139*.

[6] Brunk, E.; Rothlisberger, U. Mixed quantum mechanical/molecular mechanical molecular dynamics simulations of biological systems in ground and electronically excited states. *Chem. Rev.* **2015**, *115*, 6217–6263.

[7] Morzan, U. N.; Alonso de Armiño, D. J.; Foglia, N. O.; Ramírez, F.; González Lebrero, M. C.; Scherlis, D. A.; Estrin, D. A. Spectroscopy in Complex Environments from QM–MM Simulations. *Chem. Rev.* **2018**, *118*, 4071–4113.

[8] Gomes, A. S. P.; Jacob, C. R. Quantum-chemical embedding methods for treating local electronic excitations in complex

© The Editor(s) (if applicable) and The Author(s), under exclusive license to
Springer Fachmedien Wiesbaden GmbH, part of Springer Nature 2020
M. Scheurer, *Polarizable Embedding for the Algebraic- Diagrammatic
Construction Scheme*, BestMasters, https://doi.org/10.1007/978-3-658-31281-7

chemical systems. *Annu. Reports Sect. "C"(Physical Chem.)* **2012**, *108*, 222–277.

[9] Klamt, A.; Schüürmann, G. COSMO: a new approach to dielectric screening in solvents with explicit expressions for the screening energy and its gradient. *J. Chem. Soc., Perkin Trans. 2* **1993**, 799–805.

[10] Tomasi, J.; Mennucci, B.; Cammi, R. Quantum mechanical continuum solvation models. *Chem. Rev.* **2005**, *105*, 2999–3093.

[11] Mennucci, B. Polarizable continuum model. *Wiley Interdiscip. Rev. Comput. Mol. Sci.* **2012**, *2*, 386–404.

[12] Stamm, B.; Lagardère, L.; Scalmani, G.; Gatto, P.; Cancès, E.; Piquemal, J.-P.; Maday, Y.; Mennucci, B.; Lipparini, F. How to make continuum solvation incredibly fast in a few simple steps: a practical guide to the domain decomposition paradigm for the Conductor-like Screening Model Continuum Solvation, Linear Scaling, Domain Decomposition. *Int. J. Quantum Chem.* **2018**, in press.

[13] Mennucci, B.; Tomasi, J.; Cammi, R.; Cheeseman, J. R.; Frisch, M. J.; Devlin, F. J.; Gabriel, S.; Stephens, P. J. Polarizable Continuum Model (PCM) Calculations of Solvent Effects on Optical Rotations of Chiral Molecules. *J. Phys. Chem. A* **2002**, *106*, 6102–6113.

[14] Klein, R. A.; Mennucci, B.; Tomasi, J. Ab Initio Calculations of 17O NMR-Chemical Shifts for Water. The Limits of PCM Theory and the Role of Hydrogen-Bond Geometry and Cooperativity. *J. Phys. Chem. A* **2004**, *108*, 5851–5863.

[15] Olsen, J. M.; Aidas, K.; Kongsted, J. Excited States in Solution through Polarizable Embedding. *J. Chem. Theory Comput.* **2010**, *6*, 3721–3734.

[16] Olsen, J. M. H.; Kongsted, J. *Adv. Quantum Chem.*; Elsevier, 2011; Vol. 61; pp 107–143.

[17] Schröder, H.; Schwabe, T. Corrected Polarizable Embedding: Improving the Induction Contribution to Perichromism for Linear Response Theory. *J. Chem. Theory Comput.* **2018**, *14*, 833–842.

[18] Sneskov, K.; Schwabe, T.; Christiansen, O.; Kongsted, J. Scrutinizing the effects of polarization in QM/MM excited state calculations. *Phys. Chem. Chem. Phys.* **2011**, *13*, 18551.

[19] Sneskov, K.; Schwabe, T.; Kongsted, J.; Christiansen, O. The polarizable embedding coupled cluster method. *J. Chem. Phys.* **2011**, *134*.

[20] Schwabe, T.; Sneskov, K.; Haugaard Olsen, J. M.; Kongsted, J.; Christiansen, O.; Hättig, C. PERI-CC2: A polarizable embedded RI-CC2 method. *J. Chem. Theory Comput.* **2012**, *8*, 3274–3283.

[21] Eriksen, J. J.; Sauer, S. P.; Mikkelsen, K. V.; Jensen, H. J.; Kongsted, J. On the importance of excited state dynamic response electron correlation in polarizable embedding methods. *J. Comput. Chem.* **2012**, *33*, 2012–2022.

[22] Pedersen, M. N.; Hedegaìšrd, E. D.; Olsen, J. M. H.; Kauczor, J.; Norman, P.; Kongsted, J. Damped response theory in combination with polarizable environments: The polarizable embedding complex polarization propagator method. *J. Chem. Theory Comput.* **2014**, *10*, 1164–1171.

[23] Steindal, A. H.; Beerepoot, M. T. P.; Ringholm, M.; List, N. H.; Ruud, K.; Kongsted, J.; Olsen, J. M. H. Open-ended response theory with polarizable embedding: multiphoton absorption in biomolecular systems. *Phys. Chem. Chem. Phys.* **2016**, *18*, 28339–28352.

[24] Nåbo, L. J.; List, N. H.; Steinmann, C.; Kongsted, J. Computational Approach to Evaluation of Optical Properties of

Membrane Probes. *J. Chem. Theory Comput.* **2017**, *13*, 719–726.

[25] Nåbo, L. J.; Modzel, M.; Krishnan, K.; Covey, D. F.; Fujiwara, H.; Ory, D. S.; Szomek, M.; Khandelia, H.; Wüstner, D.; Kongsted, J. Structural design of intrinsically fluorescent oxysterols. *Chem. Phys. Lipids* **2018**, *212*, 26–34.

[26] Nørby, M. S.; Steinmann, C.; Olsen, J. M. H.; Li, H.; Kongsted, J. Computational Approach for Studying Optical Properties of DNA Systems in Solution. *J. Chem. Theory Comput.* **2016**, *12*, 5050–5057.

[27] List, N. H.; Olsen, J. M. H.; Jensen, H. J. A.; Steindal, A. H.; Kongsted, J. Molecular-level insight into the spectral tuning mechanism of the dsred chromophore. *J. Phys. Chem. Lett.* **2012**, *3*, 3513–3521.

[28] Beerepoot, M. T. P.; Steindal, A. H.; Kongsted, J.; Brandsdal, B. O.; Frediani, L.; Ruud, K.; Olsen, J. M. H. A polarizable embedding DFT study of one-photon absorption in fluorescent proteins. *Phys. Chem. Chem. Phys.* **2013**, *15*, 4735.

[29] Schwabe, T.; Beerepoot, M. T. P.; Gvan, J.; Haugaard, M.; Cd, O.; Kongsted, J. Analysis of computational models for an accurate study of electronic excitations in GFP. *Phys. Chem. Chem. Phys.* **2015**, *17*, 2582–2588.

[30] Sneskov, K.; Olsen, J. M. H.; Schwabe, T.; Hättig, C.; Christiansen, O.; Kongsted, J. Computational screening of one- and two-photon spectrally tuned channelrhodopsin mutants. *Phys. Chem. Chem. Phys.* **2013**, *15*, 7567.

[31] List, N. H.; Olsen, J. M. H.; Kongsted, J. Excited states in large molecular systems through polarizable embedding. *Phys. Chem. Chem. Phys.* **2016**, *18*, 20234–20250.

[32] Dreuw, A.; Head-Gordon, M. Failure of Time-Dependent Density Functional Theory for Long-Range Charge-Transfer Excited States: The Zincbacteriochlorin-Bacteriochlorin and Bacteriochlorophyll-Spheroidene Complexes. *J. Am. Chem. Soc.* **2004**, *126*, 4007–4016.

[33] Dreuw, A.; Head-Gordon, M. Single-reference ab initio methods for the calculation of excited states of large molecules. *Chem. Rev.* **2005**, *105*, 4009–4037.

[34] Mewes, S. A.; Plasser, F.; Dreuw, A. Communication: Exciton analysis in time-dependent density functional theory: How functionals shape excited-state characters. *J. Chem. Phys.* **2015**, *143*, 171101.

[35] Mewes, S. A.; Plasser, F.; Krylov, A.; Dreuw, A. Benchmarking Excited-State Calculations Using Exciton Properties. *J. Chem. Theory Comput.* **2018**, *14*, 710–725.

[36] Dreuw, A.; Wormit, M. The algebraic diagrammatic construction scheme for the polarization propagator for the calculation of excited states. *Wiley Interdiscip. Rev. Comput. Mol. Sci.* **2014**, *5*, 82–95.

[37] Wenzel, J.; Wormit, M.; Dreuw, A. Calculating X-ray absorption spectra of open-shell molecules with the unrestricted algebraic-diagrammatic construction scheme for the polarization propagator. *J. Chem. Theory Comput.* **2014**, *10*, 4583–4598.

[38] Wenzel, J.; Wormit, M.; Dreuw, A. Calculating core-level excitations and x-ray absorption spectra of medium-sized closed-shell molecules with the algebraic-diagrammatic construction scheme for the polarization propagator. *J. Comput. Chem.* **2014**, *35*, 1900–1915.

[39] Harbach, P. H.; Wormit, M.; Dreuw, A. The third-order algebraic diagrammatic construction method (ADC(3)) for the

polarization propagator for closed-shell molecules: Efficient implementation and benchmarking. *J. Chem. Phys.* **2014**, *141*, 064113.

[40] Mewes, J. M.; You, Z. Q.; Wormit, M.; Kriesche, T.; Herbert, J. M.; Dreuw, A. Experimental benchmark data and systematic evaluation of two a posteriori, polarizable-continuum corrections for vertical excitation energies in solution. *J. Phys. Chem. A* **2015**, *119*, 5446–5464.

[41] Mewes, J.-M.; Herbert, J. M.; Dreuw, A. On the accuracy of the general, state-specific polarizable-continuum model for the description of correlated ground- and excited states in solution. *Phys. Chem. Chem. Phys.* **2017**, *19*, 1644–1654.

[42] Lunkenheimer, B.; Köhn, A. Solvent Effects on Electronically Excited States Using the Conductor- Like Screening Model and the Second-Order Correlated Method. *J. Chem. Theory Comput.* **2013**, *9*, 977–994.

[43] Karbalaei Khani, S.; Marefat Khah, A.; Hättig, C. COSMO-RI-ADC(2) excitation energies and excited state gradients. *Phys. Chem. Chem. Phys.* **2018**, *20*, 16354–16363.

[44] Prager, S.; Zech, A.; Aquilante, F.; Dreuw, A.; Wesolowski, T. A. First time combination of frozen density embedding theory with the algebraic diagrammatic construction scheme for the polarization propagator of second order. *J. Chem. Phys.* **2016**, *144*, 204103.

[45] Prager, S.; Zech, A.; Wesolowski, T. A.; Dreuw, A. Implementation and Application of the Frozen Density Embedding Theory with the Algebraic Diagrammatic Construction Scheme for the Polarization Propagator up to Third Order. *J. Chem. Theory Comput.* **2017**, *13*, 4711–4725.

[46] Marefat Khah, A.; Karbalaei Khani, S.; Hattig, C. Analytic excited state gradients for the QM/MM polarizable embedded

second-order Algebraic Diagrammatic Construction for the polarization propagator PE-ADC (2). *Journal of chemical theory and computation* **2018**,

[47] Scheurer, M.; Brisker-Klaiman, D.; Dreuw, A. Molecular Mechanism of Flavin Photoprotection by Archaeal Dodecin: Photo-Induced Electron Transfer and Mg2+ -Promoted Proton Transfer. *J. Phys. Chem. B* **2017**, *121*, 10457–10466.

[48] Helgaker, T.; Jorgensen, P.; Olsen, J. *Molecular electronic-structure theory*; John Wiley & Sons, 2014.

[49] Schirmer, J. Beyond the random-phase approximation: A new approximation scheme for the polarization propagator. *Phys. Rev. A* **1982**, *26*, 2395–2416.

[50] Schirmer, J.; Trofimov, A. B. Intermediate state representation approach to physical properties of electronically excited molecules. *J. Chem. Phys.* **2004**, *120*, 11449–11464.

[51] Wormit, M.; Rehn, D. R.; Harbach, P. H.; Wenzel, J.; Krauter, C. M.; Epifanovsky, E.; Dreuw, A. Investigating excited electronic states using the algebraic diagrammatic construction (ADC) approach of the polarisation propagator. *Mol. Phys.* **2014**, *112*, 774–784.

[52] Trofimov, A. B.; Schirmer, J. An efficient polarization propagator approach to valence electron excitation spectra. *J. Phys. B At. Mol. Opt. Phys.* **1995**, *28*, 2299–2324.

[53] Applequist, J.; Carl, J. R.; Fung, K.-K. Atom dipole interaction model for molecular polarizability. Application to polyatomic molecules and determination of atom polarizabilities. *J. Am. Chem. Soc.* **1972**, *94*, 2952–2960.

[54] Day, P. N.; Jensen, J. H.; Gordon, M. S.; Webb, S. P.; Stevens, W. J.; Krauss, M.; Garmer, D.; Basch, H.; Cohen, D. An effective fragment method for modeling solvent effects in

quantum mechanical calculations. *J. Chem. Phys.* **1996**, *105*, 1968–1986.

[55] Gordon, M. S.; Freitag, M. A.; Bandyopadhyay, P.; Jensen, J. H.; Kairys, V.; Stevens, W. J. The Effective Fragment Potential Method: A QM-Based MM Approach to Modeling Environmental Effects in Chemistry. *J. Phys. Chem. A* **2001**, *105*, 293–307.

[56] Gordon, M. S.; Slipchenko, L.; Li, H.; Jensen, J. H. The Effective Fragment Potential: A General Method for Predicting Intermolecular Interactions. *Annu. Rep. Comput. Chem.* **2007**, *3*, 177–193.

[57] Gordon, M. S.; Mullin, J. M.; Pruitt, S. R.; Roskop, L. B.; Slipchenko, L. V.; Boatz, J. A. Accurate methods for large molecular systems. *J. Phys. Chem. B* **2009**, *113*, 9646–9663.

[58] Ghosh, D.; Kosenkov, D.; Vanovschi, V.; Williams, C. F.; Herbert, J. M.; Gordon, M. S.; Schmidt, M. W.; Slipchenko, L. V.; Krylov, A. I. Noncovalent interactions in extended systems described by the effective fragment potential method: Theory and application to nucleobase oligomers. *J. Phys. Chem. A* **2010**, *114*, 12739–12754.

[59] Slipchenko, L. V. Solvation of the excited states of chromophores in polarizable environment: Orbital relaxation versus polarization. *J. Phys. Chem. A* **2010**, *114*, 8824–8830.

[60] Kosenkov, D.; Slipchenko, L. V. Solvent Effects on the Electronic Transitions of p -Nitroaniline : A QM / EFP Study. *J. Phys. Chem. A* **2010**, *115*, 392–401.

[61] Schwabe, T. General theory for environmental effects on (vertical) electronic excitation energies. *J. Chem. Phys.* **2016**, *145*, 154105.

[62] Caricato, M.; Mennucci, B.; Tomasi, J.; Ingrosso, F.; Cammi, R.; Corni, S.; Caricato, M.; Mennucci, B.; Tomasi, J. Formation and

relaxation of excited states in solution : A new time dependent polarizable continuum model based on time dependent density functional theory. *J. Chem. Phys.* **2006**, *124*, 124520.

[63] Aidas, K.; Angeli, C.; Bak, K. L.; Bakken, V.; Bast, R.; Boman, L.; Christiansen, O.; Cimiraglia, R.; Coriani, S.; Dahle, P.; Dalskov, E. K.; Ekström, U.; Enevoldsen, T.; Eriksen, J. J.; Ettenhuber, P.; Fernández, B.; Ferrighi, L.; Fliegl, H.; Frediani, L.; Hald, K.; Halkier, A.; Hättig, C.; Heiberg, H.; Helgaker, T.; Hennum, A. C.; Hettema, H.; Hjertenæs, E.; Høst, S.; Høyvik, I.-M.; Iozzi, M. F.; Jansík, B.; Jensen, H. J. Aa.; Jonsson, D.; Jørgensen, P.; Kauczor, J.; Kirpekar, S.; Kjærgaard, T.; Klopper, W.; Knecht, S.; Kobayashi, R.; Koch, H.; Kongsted, J.; Krapp, A.; Kristensen, K.; Ligabue, A.; Lutnæs, O. B.; Melo, J. I.; Mikkelsen, K. V.; Myhre, R. H.; Neiss, C.; Nielsen, C. B.; Norman, P.; Olsen, J.; Olsen, J. M. H.; Osted, A.; Packer, M. J.; Pawlowski, F.; Pedersen, T. B.; Provasi, P. F.; Reine, S.; Rinkevicius, Z.; Ruden, T. A.; Ruud, K.; Rybkin, V. V.; Sałek, P.; Samson, C. C. M.; de Merás, A. S.; Saue, T.; Sauer, S. P. A.; Schimmelpfennig, B.; Sneskov, K.; Steindal, A. H.; Sylvester-Hvid, K. O.; Taylor, P. R.; Teale, A. M.; Tellgren, E. I.; Tew, D. P.; Thorvaldsen, A. J.; Thøgersen, L.; Vahtras, O.; Watson, M. A.; Wilson, D. J. D.; Ziolkowski, M.; Ågren, H. The Dalton quantum chemistry program system. *WIREs Comput. Mol. Sci.* **2014**, *4*, 269–284.

[64] Smith, D. G.; Burns, L. A.; Sirianni, D. A.; Nascimento, D. R.; Kumar, A.; James, A. M.; Schriber, J. B.; Zhang, T.; Zhang, B.; Abbott, A. S. Psi4NumPy: An Interactive Quantum Chemistry Programming Environment for Reference Implementations and Rapid Development. *Journal of chemical theory and computation* **2018**,

[65] Sun, Q.; Berkelbach, T. C.; Blunt, N. S.; Booth, G. H.; Guo, S.; Li, Z.; Liu, J.; McClain, J. D.; Sayfutyarova, E. R.; Sharma, S. PySCF: the Python-based simulations of chemistry framework.

Wiley Interdisciplinary Reviews: Computational Molecular Science **2018**, *8*, e1340.

[66] Rinkevicius, Z.; Li, X.; Vahtras, O.; Ahmadzadeh, K.; Brand, M.; Ringholm, M.; List, N. H.; Scheurer, M.; Scott, M.; Dreuw, A.; Norman, P. VeloxChem: A Python-driven density-functional theory program for spectroscopy simulations in high-performance computing environments. *WIREs Comput. Mol. Sci.* e1457.

[67] Scheurer, M.; Reinholdt, P.; Kjellgren, E. R.; Olsen, J. M. H.; Dreuw, A.; Kongsted, J. CPPE: An Open-Source C++ and Python Library for Polarizable Embedding. *J. Chem. Theory Comput.* **2019**, *15*, 6154–6163.

[68] Shao, Y.; Gan, Z.; Epifanovsky, E.; Gilbert, A. T.; Wormit, M.; Kussmann, J.; Lange, A. W.; Behn, A.; Deng, J.; Feng, X.; Ghosh, D.; Goldey, M.; Horn, P. R.; Jacobson, L. D.; Kaliman, I.; Khaliullin, R. Z.; Kuś, T.; Landau, A.; Liu, J.; Proynov, E. I.; Rhee, Y. M.; Richard, R. M.; Rohrdanz, M. A.; Steele, R. P.; Sundstrom, E. J.; Woodcock, H. L.; Zimmerman, P. M.; Zuev, D.; Albrecht, B.; Alguire, E.; Austin, B.; Beran, G. J. O.; Bernard, Y. A.; Berquist, E.; Brandhorst, K.; Bravaya, K. B.; Brown, S. T.; Casanova, D.; Chang, C.-M.; Chen, Y.; Chien, S. H.; Closser, K. D.; Crittenden, D. L.; Diedenhofen, M.; DiStasio, R. A.; Do, H.; Dutoi, A. D.; Edgar, R. G.; Fatehi, S.; Fusti-Molnar, L.; Ghysels, A.; Golubeva-Zadorozhnaya, A.; Gomes, J.; Hanson-Heine, M. W.; Harbach, P. H.; Hauser, A. W.; Hohenstein, E. G.; Holden, Z. C.; Jagau, T.-C.; Ji, H.; Kaduk, B.; Khistyaev, K.; Kim, J.; Kim, J.; King, R. A.; Klunzinger, P.; Kosenkov, D.; Kowalczyk, T.; Krauter, C. M.; Lao, K. U.; Laurent, A. D.; Lawler, K. V.; Levchenko, S. V.; Lin, C. Y.; Liu, F.; Livshits, E.; Lochan, R. C.; Luenser, A.; Manohar, P.; Manzer, S. F.; Mao, S.-P.; Mardirossian, N.; Marenich, A. V.; Maurer, S. A.; Mayhall, N. J.; Neuscamman, E.; Oana, C. M.; Olivares-Amaya, R.; O'Neill, D. P.; Parkhill, J. A.; Perrine, T. M.; Peverati, R.;

Prociuk, A.; Rehn, D. R.; Rosta, E.; Russ, N. J.; Sharada, S. M.; Sharma, S.; Small, D. W.; Sodt, A.; Stein, T.; Stück, D.; Su, Y.-C.; Thom, A. J.; Tsuchimochi, T.; Vanovschi, V.; Vogt, L.; Vydrov, O.; Wang, T.; Watson, M. A.; Wenzel, J.; White, A.; Williams, C. F.; Yang, J.; Yeganeh, S.; Yost, S. R.; You, Z.-Q.; Zhang, I. Y.; Zhang, X.; Zhao, Y.; Brooks, B. R.; Chan, G. K.; Chipman, D. M.; Cramer, C. J.; Goddard, W. A.; Gordon, M. S.; Hehre, W. J.; Klamt, A.; Schaefer, H. F.; Schmidt, M. W.; Sherrill, C. D.; Truhlar, D. G.; Warshel, A.; Xu, X.; Aspuru-Guzik, A.; Baer, R.; Bell, A. T.; Besley, N. A.; Chai, J.-D.; Dreuw, A.; Dunietz, B. D.; Furlani, T. R.; Gwaltney, S. R.; Hsu, C.-P.; Jung, Y.; Kong, J.; Lambrecht, D. S.; Liang, W.; Ochsenfeld, C.; Rassolov, V. A.; Slipchenko, L. V.; Subotnik, J. E.; Van Voorhis, T.; Herbert, J. M.; Krylov, A. I.; Gill, P. M.; Head-Gordon, M. Advances in molecular quantum chemistry contained in the Q-Chem 4 program package. *Mol. Phys.* **2015**, *113*, 184–215.

[69] Neese, F. Software update: the ORCA program system, version 4.0. *Wiley Interdiscip. Rev. Comput. Mol. Sci.* **2018**, *8*.

[70] Herbst, M. F.; Dreuw, A.; Avery, J. E. Towards quantum-chemical method development for arbitrary basis functions. *arXiv preprint arXiv:1807.00704* **2018**,

[71] Behnel, S.; Bradshaw, R.; Citro, C.; Dalcin, L.; Seljebotn, D. S.; Smith, K. Cython: The Best of Both Worlds. *Comput. Sci. Eng.* **2011**, *13*, 31–39.

[72] Beazley, D. M. SWIG: An Easy to Use Tool for Integrating Scripting Languages with C and C++. Tcl/Tk Workshop. 1996.

[73] Rick, S. W.; Stuart, S. J.; Berne, B. J. Dynamical fluctuating charge force fields: Application to liquid water. *The Journal of chemical physics* **1994**, *101*, 6141–6156.

[74] Sanderson, C.; Curtin, R. Armadillo: a template-based C++ library for linear algebra. *Journal of Open Source Software* **2016**,

[75] Trofimov, A. B.; Stelter, G.; Schirmer, J. A consistent third-order propagator method for electronic excitation. *J. Chem. Phys.* **1999**, *111*, 9982–9999.

[76] Hermann, G.; Pohl, V.; Tremblay, J. C.; Paulus, B.; Hege, H. C.; Schild, A. ORBKIT: A modular python toolbox for cross-platform postprocessing of quantum chemical wavefunction data. *J. Comput. Chem.* **2016**, *37*, 1511–1520.

[77] McKinney, W. Data Structures for Statistical Computing in Python. Proc. 9th Python Sci. Conf. 2010; pp 51–56.

[78] McKinney, W. pandas: a Foundational Python Library for Data Analysis and Statistics. 2011.

[79] Van Der Walt, S.; Colbert, S. C.; Varoquaux, G. The NumPy array: A structure for efficient numerical computation. *Comput. Sci. Eng.* **2011**, *13*, 22–30.

[80] Oliphant, T. E. SciPy: Open source scientific tools for Python. *Comput. Sci. Eng.* **2007**, *9*, 10–20.

[81] Hunter, J. D. Matplotlib: A 2D Graphics Environment. *Comput. Sci. Eng.* **2007**, *9*, 99–104.

[82] Waskom, M.; Botvinnik, O.; O'Kane, D.; Hobson, P.; Lukauskas, S.; Gemperline, D. C.; Augspurger, T.; Halchenko, Y.; Cole, J. B.; Warmenhoven, J.; de Ruiter, J.; Pye, C.; Hoyer, S.; Vanderplas, J.; Villalba, S.; Kunter, G.; Quintero, E.; Bachant, P.; Martin, M.; Meyer, K.; Miles, A.; Ram, Y.; Yarkoni, T.; Williams, M. L.; Evans, C.; Fitzgerald, C.; Brian,; Fonnesbeck, C.; Lee, A.; Qalieh, A. Mwaskom/Seaborn: V0.8.1. 2017; https://doi.org/10.5281/zenodo.883859.

[83] Dunning, T. H. Gaussian basis sets for use in correlated molecular calculations. I. The atoms boron through neon and hydrogen. *J. Chem. Phys.* **1989**, *90*, 1007–1023.

[84] Kendall, R. A.; Dunning, T. H.; Harrison, R. J. Electron affinities of the first-row atoms revisited. Systematic basis sets and wave functions. *J. Chem. Phys.* **1992**, *96*, 6796–6806.

[85] Woon, D. E.; Dunning, T. H. Gaussian basis sets for use in correlated molecular calculations. III. The atoms aluminum through argon. *J. Chem. Phys.* **1993**, *98*, 1358–1371.

[86] Olsen, J. M. H. PyFraME: Python tools for Fragment-based Multiscale Embedding (version 0.1.1). 2018; https://doi.org/10.5281/zenodo.1168860.

[87] Gagliardi, L.; Lindh, R.; Karlström, G. Local properties of quantum chemical systems: The LoProp approach. *J. Chem. Phys.* **2004**, *121*, 4494–4500.

[88] Adamo, C.; Barone, V. Toward reliable density functional methods without adjustable parameters: The PBE0 model. *J. Chem. Phys.* **1999**, *110*, 6158–6170.

[89] Vahtras, O. LoProp for Dalton. 2014; https://doi.org/10.5281/zenodo.13276.

[90] Plasser, F.; Wormit, M.; Dreuw, A. New tools for the systematic analysis and visualization of electronic excitations. I. Formalism. *J. Chem. Phys.* **2014**, *141*, 024106.

[91] Plasser, F.; Bäppler, S. A.; Wormit, M.; Dreuw, A. New tools for the systematic analysis and visualization of electronic excitations. II. Applications. *J. Chem. Phys.* **2014**, *141*, 024107.

[92] Humphrey, W.; Dalke, A.; Schulten, K. VMD: Visual molecular dynamics. *J. Mol. Graph.* **1996**, *14*, 33–38.

[93] Sure, R.; Grimme, S. Corrected small basis set Hartree-Fock method for large systems. *J. Comput. Chem.* **2013**, *34*, 1672–1685.

[94] Neese, F. The ORCA program system. *Wiley Interdiscip. Rev. Comput. Mol. Sci.* **2012**, *2*, 73–78.

[95] Martinez, L.; Andrade, R.; Birgin, E. G.; Martínez, J. M. PACK-MOL: A package for building initial configurations for molecular dynamics simulations. *J. Comput. Chem.* **2009**, *30*, 2157–2164.

[96] Phillips, J. C.; Braun, R.; Wang, W.; Gumbart, J.; Tajkhorshid, E.; Villa, E.; Chipot, C.; Skeel, R. D.; Kalé, L.; Schulten, K. Scalable molecular dynamics with NAMD. *J. Comput. Chem.* **2005**, *26*, 1781–1802.

[97] Huang, J.; Mackerell, A. D. CHARMM36 all-atom additive protein force field: Validation based on comparison to NMR data. *J. Comput. Chem.* **2013**, *34*, 2135–2145.

[98] Zoete, V.; Cuendet, M. A.; Grosdidier, A.; Michielin, O. SwissParam: A fast force field generation tool for small organic molecules. *J. Comput. Chem.* **2011**, *32*, 2359–2368.

[99] Martyna, G. J.; Tobias, D. J.; Klein, M. L. Constant pressure molecular dynamics algorithms. *The Journal of Chemical Physics* **1994**, *101*, 4177–4189.

[100] Feller, S. E.; Zhang, Y.; Pastor, R. W.; Brooks, B. R. Constant pressure molecular dynamics simulation: the Langevin piston method. *The Journal of chemical physics* **1995**, *103*, 4613–4621.

[101] Miyamoto, S.; Kollman, P. A. Settle: An analytical version of the SHAKE and RATTLE algorithm for rigid water models. *J. Comput. Chem.* **1992**, *13*, 952–962.

[102] Andersen, H. C. Rattle: A "velocity"version of the shake algorithm for molecular dynamics calculations. *J. Comput. Phys.* **1983**, *52*, 24–34.

[103] Darden, T.; York, D.; Pedersen, L. Particle mesh Ewald: An N log(N) method for Ewald sums in large systems. *J. Chem. Phys.* **1993**, *98*, 10089–10092.

[104] Melo, M. C. R.; Bernardi, R. C.; Rudack, T.; Scheurer, M.; Riplinger, C.; Phillips, J. C.; Maia Julio D C, R. G. B.; Ribeiro, J. V.; Stone, J. E.; Neese, F.; Schulten, K.; Luthey-Schulten, Z. NAMD goes quantum: An integrative suite for hybrid simulations. *Nat. Methods* **2018**, *15*, 351–354.

[105] Michaud-Agrawal, N.; Denning, E. J.; Woolf, T. B.; Beckstein, O. MDAnalysis: a toolkit for the analysis of molecular dynamics simulations. *J. Comput. Chem.* **2011**, *32*, 2319–2327.

[106] Gowers, R. J.; Linke, M.; Barnoud, J.; Reddy, T. J. E.; Melo, M. N.; Seyler, S. L.; Domański, J.; Dotson, D. L.; Buchoux, S.; Kenney, I. M.; Beckstein, O. MDAnalysis: A Python Package for the Rapid Analysis of Molecular Dynamics Simulations. *Proc. 15th Python Sci. Conf.* **2016**, 98–105.

[107] Beerepoot, M. T.; Steindal, A. H.; Ruud, K.; Olsen, J. M. H.; Kongsted, J. Convergence of environment polarization effects in multiscale modeling of excitation energies. *Comput. Theor. Chem.* **2014**, *1040-1041*, 304–311.

[108] Sherwood, P.; H. de Vries, A.; J. Collins, S.; P. Greatbanks, S.; A. Burton, N.; A. Vincent, M.; H. Hillier, I. Computer simulation of zeolite structure and reactivity using embedded cluster methods. *Faraday Discuss.* **1997**, *106*, 79–92.

[109] Becke, A. D. Density-functional thermochemistry. III. The role of exact exchange. *The Journal of chemical physics* **1993**, *98*, 5648–5652.

[110] Beerepoot, M. T.; Steindal, A. H.; List, N. H.; Kongsted, J.; Olsen, J. M. H. Averaged Solvent Embedding Potential Parameters for Multiscale Modeling of Molecular Properties. *Journal of chemical theory and computation* **2016**, *12*, 1684–1695.

[111] Steinmann, C.; Reinholdt, P.; Nørby, M. S.; Kongsted, J.; Olsen, J. M. H. Response properties of embedded mo-

lecules through the polarizable embedding model. **2018**, https://arxiv.org/abs/1804.03598.

[112] Yang, C.; Dreuw, A. Evaluation of the restricted virtual space approximation in the algebraic-diagrammatic construction scheme for the polarization propagator to speed-up excited-state calculations. *J. Comput. Chem.* **2017**, *38*, 1528–1537.

[113] Scheurer, M.; Herbst, M. F.; Reinholdt, P.; Olsen, J. M. H.; Dreuw, A.; Kongsted, J. Polarizable Embedding Combined with the Algebraic Diagrammatic Construction: Tackling Excited States in Biomolecular Systems. *J. Chem. Theory Comput.* **2018**, *14*, 4870–4883.

[114] Reinholdt, P.; Kongsted, J.; Olsen, J. M. H. Polarizable Density Embedding: A Solution to the Electron Spill-Out Problem in Multiscale Modeling. *J. Phys. Chem. Lett.*

[115] Kongsted, J.; Osted, A.; Mikkelsen, K. V.; Åstrand, P. O.; Christiansen, O. Solvent effects on the n π^* electronic transition in formaldehyde: A combined coupled cluster/molecular dynamics study. *J. Chem. Phys.* **2004**, *121*, 8435–8445.

[116] Kongsted, J.; Osted, A.; Pedersen, T. B.; Mikkelsen, K. V.; Christiansen, O. The n π^* electronic transition in microsolvated formaldehyde. A coupled cluster and combined coupled cluster/molecular mechanics study. *J. Phys. Chem. A* **2004**, *108*, 8624–8632.

[117] Thomsen, C. L.; Thøgersen, J.; Keiding, S. R. Ultrafast Charge-Transfer Dynamics: Studies of p -Nitroaniline in Water and Dioxane. *J. Phys. Chem. A* **1998**, *102*, 1062–1067.

[118] Kovalenko, S.; Schanz, R.; Farztdinov, V.; Hennig, H.; Ernsting, N. Femtosecond relaxation of photoexcited para-nitroaniline: solvation, charge transfer, internal conversion and cooling. *Chem. Phys. Lett.* **2000**, *323*, 312–322.

[119] Hättig, C. Structure optimizations for excited states with correlated second-order methods: CC2 and ADC (2). *Adv. Quantum Chem.* **2005**, *50*, 37–60.

[120] Grininger, M.; Staudt, H.; Johansson, P.; Wachtveitl, J.; Oesterhelt, D. Dodecin is the key player in flavin homeostasis of archaea. *J. Biol. Chem.* **2009**, *284*, 13068–13076.

[121] Staudt, H.; Oesterhelt, D.; Grininger, M.; Wachtveitl, J. Ultrafast excited-state deactivation of flavins bound to dodecin. *J. Biol. Chem.* **2012**, *287*, 17637–17644.

[122] Staudt, H.; Hoesl, M. G.; Dreuw, A.; Serdjukow, S.; Oesterhelt, D.; Budisa, N.; Wachtveitl, J.; Grininger, M. Directed manipulation of a flavoprotein photocycle. *Angew. Chemie - Int. Ed.* **2013**, *52*, 8463–8466.

[123] Hättig, C. Structure optimizations for excited states with correlated second-order methods: CC2 and ADC (2). *Adv. Quantum Chem.* **2005**, *50*, 37–60.

[124] Wormit, M. Development and application of reliable methods for the calculation of excited states : from light-harvesting complexes to medium-sized molecules. Ph.D. thesis, 2009.

[125] Fransson, T.; Harada, Y.; Kosugi, N.; Besley, N. A.; Winter, B.; Rehr, J. J.; Pettersson, L. G.; Nilsson, A. X-ray and electron spectroscopy of water. *Chem. Rev.* **2016**, *116*, 7551–7569.

[126] Zhovtobriukh, I.; Besley, N. A.; Fransson, T.; Nilsson, A.; Pettersson, L. G. Relationship between x-ray emission and absorption spectroscopy and the local H-bond environment in water. *J. Chem. Phys.* **2018**, *148*, 144507.

[127] Harbach, P. H.; Schneider, M.; Faraji, S.; Dreuw, A. Intermolecular Coulombic Decay in Biology: The Initial Electron Detachment from FADH–in DNA Photolyases. *J. Phys. Chem. Lett.* **2013**, *4*, 943–949.

[128] He, T. F.; Guo, L.; Guo, X.; Chang, C. W.; Wang, L.; Zhong, D. Femtosecond dynamics of short-range protein electron transfer in flavodoxin. *Biochemistry* **2013**, *52*, 9120–9128.

[129] Lee, J.; Head-Gordon, M. Regularized Orbital-Optimized Second-Order Møller-Plesset Perturbation Theory: A Reliable Fifth-Order-Scaling Electron Correlation Model with Orbital Energy Dependent Regularizers. *J. Chem. Theor. Comput.* **2018**, *10.1021/acs.jctc.8b00731*, Article ASAP.

A Original Implementation of PE in Q-Chem

The first pilot implementation of CPPE was done with focus on the *Q-Chem* program package for electronic structure calculations. The CPPE library has since been redesigned and rewritten to be more general such that integration to other program packages is facilitated. In the following, the original interface designed for *Q-Chem* is outlined.

A.1 The PeCalcHandler Class

The upmost level of CPPE that needs to be interfaced with the host program resides in the abstract base class called PeCalcHandler. As the name suggests, the class is responsible to handle a PE calculation and expose the needed functionality to the host program. This class only contains pure virtual methods, such that it needs to be subclassed and the methods be defined inside the subclasses. For each back-end, one subclass is necessary. The methods comprise initialization (bool initialize()), fock contribution (void fock_contribution(...)), printing the energy (void print_energy()), and perturbative energy corrections (void perturbative_exc_energy_correction(...)). Upon host program initialization and option read-in, the back-end decision is made, a new instance of a subclass of PeCalcHandler is created, and a pointer to the abstract class is stored. Further, the input parsing creates a container for PE calculation options (e.g., potential file name), libcppe::PeOptions, that is subsequently passed to the calculation handlers. Using the stored pointer, calls to the methods of the respective implementation in the subclass are feasible. Thus, this concept of *dynamic linkage* allows the program to decide on runtime which code should be executed based on the kind of object for which a method is called. Until now, CPPE has been interfaced with the *Q-Chem* program package, and the two created subclasses of `PeCalcHandler` are called `PelibPeCalcHandler` and `CppePeCalcHandler` for the `PElib` and the CPPE-internal back-end, respectively. The implementations of the above methods reside in the host program as they require access to host-program-specific routines and data, such as the core molecule, the AO basis or the integral engine.

A.2 Interface to `PElib`

In the pilot version of CPPE, an interface to the existing `PElib`, which contains the original implementation of the PE model by Olsen and Kongsted, was created for testing purposes. Since CPPE is now properly tested and working, the interface has become obsolete.

The interface to the `PElib` `FORTRAN` code is rather simple. The only difficulty is in appropriately passing data from C++ to `FORTRAN`, which is only feasible using pointers to respective memory locations. In that manner, all needed routines in `PElib` and `gen1int` are wrapped by `extern C` functions (`cppe/cppe/interface.h`). These C routines are again wrapped in `cppe/cppe/libcppe.cc` and can then be called by the `PelibPeCalcHandler` methods:

- initialize `PElib` and `gen1int`
  (`PeLibPeCalcHandler::initialize()`)

  - obtain molecular geometry and nuclear charges

  - pass AO basis from *Q-Chem* to `gen1int` by shell

  - pass PE options to `PElib`

- Fock contribution
  (`PeLibPeCalcHandler::fock_contribution(...)`)

  - provided density matrix (`arma::mat Ptot`) is folded, and the pointer passed to `PElib`, together with memory location for results

  - `PElib` calculates the embedding energy and operator

- embedding energy and operator are passed to *Q-Chem*

- Energy print-out
  (PelibPeCalcHandler::print_energy())
  - PElib calculates and prints the embedding energy summary for a given density matrix

- Perturbative correction
  (perturbative_exc_energy_correction(...))
  - PElib calculates and returns the electronic induction energy based on a given density matrix

To summarize, most work in programming the interface lies in correctly initializing PElib and gen1int. For the latter, the AO shells need to be passed to FORTRAN. I have therefore un-entangled the gen1int dependencies in *Dalton*, and written a new FORTRAN routine inside gen1int to easily pass shell data over to the library. Wrapping of matrices received from *Q-Chem* is straightforward since PElib is henceforth used in a 'black-box' manner.

A.3 Interface to CPPE-internal Back-End

Even though the overall structure of the calculation handler for CPPE-internal PE calculations (CppePeCalcHandler) is similar to PeLibPeCalcHandler, the implementation is slightly more involved since the required integrals need to be obtained from the host program. In addition, the core and utils code in CPPE provide the necessary building blocks and routines for potential file read-in as well as calculation of embedding energies and operators. More details can be found in the next section. Again, the implemented methods are described:

- initialize CPPE (CppePeCalcHandler::initialize())
  - obtain molecular geometry and nuclear charges
  - CPPE reads the potential file (using PotfileReader)

- potentials are modified (using `PotManipulator`) according to `PeOptions`

- Fock contribution
  (`CppePeCalcHandler::fock_contribution(...)`)

  - integrals for electrostatics are calculated ($t_{pq}^{(k)}$) using `libqints` and stored in memory

  - static interaction energies and fields are calculated

  - $\underline{\mathbf{F}}_{\text{el}}$ are calculated by contracting the field integrals with the density matrix

  - induced moments are obtained (see 4.2.1)

  - induction operator is calculated

  - PE operator is added to the *Q-Chem* Fock matrix

- Energy print-out
  (`CppePeCalcHandler::print_energy()`)

  - CPPE prints a PE energy summary to the output

- Perturbative correction
  (`perturbative_exc_energy_correction(...)`)

  - $\underline{\mathbf{F}}_{\text{el}}$ are calculated by contracting the field integrals with the given density matrix

  - induced moments based on $\underline{\mathbf{F}}_{\text{el}}$ are calculated

  - induction energy is calculated

B Recasting the Interaction Hamiltonian

The expressions for all Taylor expansions of the Coulomb interaction operator are given by

$$
\frac{1}{|\mathbf{r} - \mathbf{r}'|} = \sum_{|k|=0}^{\infty} \frac{(-1)^{|k|}}{k!} \left(\nabla^k \frac{1}{|\mathbf{r} - \underline{\mathbf{R}}_o|} \right) \left(\mathbf{r}' - \underline{\mathbf{R}}_o \right)^k
$$

$$
= \sum_{|k|=0}^{\infty} \frac{(-1)^{|k|}}{k!} T_{\mathrm{AB}}^{(k)}(\mathbf{r}) \left(\mathbf{r}' - \underline{\mathbf{R}}_o \right)^k \tag{B.1}
$$

$$
\frac{1}{|\mathbf{r} - \underline{\mathbf{R}}_m|} = \sum_{|k|=0}^{\infty} \frac{(-1)^{|k|}}{k!} T_{\mathrm{AB}}^{(k)}(\mathbf{r}) (\underline{\mathbf{R}}_m - \underline{\mathbf{R}}_o)^k \tag{B.2}
$$

$$
\frac{1}{|\underline{\mathbf{R}}_n - \mathbf{r}'|} = \sum_{|k|=0}^{\infty} \frac{(-1)^{|k|}}{k!} T_{\mathrm{AB}}^{(k)}(\underline{\mathbf{R}}_n) \left(\mathbf{r}' - \underline{\mathbf{R}}_o \right)^k \tag{B.3}
$$

$$
\frac{1}{|\underline{\mathbf{R}}_n - \underline{\mathbf{R}}_m|} = \sum_{|k|=0}^{\infty} \frac{(-1)^{|k|}}{k!} T_{\mathrm{AB}}^{(k)}(\underline{\mathbf{R}}_n)(\underline{\mathbf{R}}_m - \underline{\mathbf{R}}_o)^k \tag{B.4}
$$

Using these expressions, we can write the interaction Hamiltonian as

$$\hat{V}^{AB} = \sum_{|k|=0}^{\infty} \frac{(-1)^{|k|}}{k!} \underbrace{\sum_{m=1}^{M_B} Z_m^B \left(\underline{\mathbf{R}}_m - \underline{\mathbf{R}}_o\right)^k}_{Q_{B,nuc}^{(k)}}$$

$$\underbrace{\sum_{pq \in A} \left(-\int \phi_p^*(\underline{\mathbf{r}}) T_{AB}^{(k)}(\underline{\mathbf{r}}) \phi_q(\underline{\mathbf{r}}) d\underline{\mathbf{r}} \right) \hat{E}_{pq}^A}_{\hat{\mathcal{V}}_{A,el}^{(k)}}$$

$$+ \sum_{|k|=0}^{\infty} \frac{(-1)^{|k|}}{k!} \underbrace{\sum_{n=1}^{M_A} Z_n^A T_{AB}^{(k)}(\underline{\mathbf{R}}_n)}_{\mathcal{V}_{A,nuc}^{(k)}}$$

$$\underbrace{\sum_{rs \in B} \left(-\int \phi_r^*(\underline{\mathbf{r}'}) \left(\underline{\mathbf{r}'} - \underline{\mathbf{R}}_o\right)^k \phi_s(\underline{\mathbf{r}'}) d\underline{\mathbf{r}'} \right) \hat{E}_{rs}^B}_{\hat{\mathcal{Q}}_{B,el}^{(k)}}$$

$$+ \sum_{|k|=0}^{\infty} \frac{(-1)^{|k|}}{k!} \underbrace{\sum_{pq \in A} \left(-\int \phi_p^*(\underline{\mathbf{r}}) T_{AB}^{(k)}(\underline{\mathbf{r}}) \phi_q(\underline{\mathbf{r}}) d\underline{\mathbf{r}} \right) \hat{E}_{pq}^A}_{\hat{\mathcal{V}}_{A,el}^{(k)}}$$

$$\underbrace{\sum_{rs \in B} \left(-\int \phi_r^*(\underline{\mathbf{r}'}) \left(\underline{\mathbf{r}'} - \underline{\mathbf{R}}_o\right)^k \phi_s(\underline{\mathbf{r}'}) d\underline{\mathbf{r}'} \right) \hat{E}_{rs}^B}_{\hat{\mathcal{Q}}_{B,el}^{(k)}}$$

$$+ \sum_{|k|=0}^{\infty} \frac{(-1)^{|k|}}{k!} \underbrace{\sum_{n=1}^{M_A} Z_n^A T_{AB}^{(k)}(\underline{\mathbf{R}}_n)}_{\mathcal{V}_{A,nuc}^{(k)}} \underbrace{\sum_{m=1}^{M_B} Z_m^B \left(\underline{\mathbf{R}}_m - \underline{\mathbf{R}}_o\right)^k}_{\mathcal{Q}_{B,nuc}^{(k)}}. \quad (B.5)$$

Collecting terms in a fragment-wise manner, we arrive at

$$\hat{V}^{\mathrm{AB}} = \sum_{|k|=0}^{\infty} \frac{(-1)^{|k|}}{k!} \left(\hat{\mathcal{V}}_{\mathrm{A,nuc}}^{(k)} + \hat{\mathcal{V}}_{\mathrm{A,el}}^{(k)} \right) \left(\hat{\mathcal{Q}}_{\mathrm{B,nuc}}^{(k)} + \hat{\mathcal{Q}}_{\mathrm{B,el}}^{(k)} \right) \qquad \text{(B.6)}$$

$$= \sum_{|k|=0}^{\infty} \frac{(-1)^{|k|}}{k!} \, \hat{\mathcal{V}}_{\mathrm{A}}^{(k)} \hat{\mathcal{Q}}_{\mathrm{B}}^{(k)} \, . \qquad \text{(B.7)}$$

Publication List

1. Polarizable Embedding Combined with the Algebraic Diagrammatic Construction: Tackling Excited States in Biomolecular Systems. **Maximilian Scheurer**, Michael F. Herbst, Peter Reinholdt, Jógvan Magnus Haugaard Olsen, Andreas Dreuw, and Jacob Kongsted
   J. Chem. Theor. Comput., **2018**, 14 (9), 4870-4883.

2. CPPE: An Open-Source C++ and Python Library for Polarizable Embedding. **Maximilian Scheurer**, Peter Reinholdt, Erik Rosendahl Kjellgren, Jógvan Magnus Haugaard Olsen, Andreas Dreuw, and Jacob Kongsted
   J. Chem. Theor. Comput., **2019**, 15 (11), 6154-6163.

Printed in the United States
By Bookmasters